Surroundings

Surroundings

A HISTORY OF ENVIRONMENTS
AND ENVIRONMENTALISMS

Etienne S. Benson

The University of Chicago Press   CHICAGO & LONDON

The University of Chicago Press, Chicago 60637
The University of Chicago Press, Ltd., London
© 2020 by The University of Chicago
Published 2020
Printed in the United States of America

29 28 27 26 25 24 23 22 21 20   1 2 3 4 5

ISBN-13: 978-0-226-70615-3 (cloth)
ISBN-13: 978-0-226-70629-0 (paper)
ISBN-13: 978-0-226-70632-0 (e-book)
DOI: https://doi.org/10.7208/chicago/9780226706320.001.0001

Library of Congress Cataloging-in-Publication Data

Names: Benson, Etienne, 1976– author.
Title: Surroundings : a history of environments and environmentalisms / Etienne S. Benson.
Description: Chicago : University of Chicago Press, 2020. | Includes bibliographical
  references and index.
Identifiers: LCCN 2019042712 | ISBN 9780226706153 (cloth) | ISBN 9780226706290
  (paperback) | ISBN 9780226706320 (ebook)
Subjects: LCSH: Environmentalism—History. | Environmentalism—United States—History.
Classification: LCC GE195.B4638 2020 | DDC 363.700973—dc23
LC record available at https://lccn.loc.gov/2019042712

♾ This paper meets the requirements of ANSI/NISO Z39.48-1992 (Permanence of Paper).

CONTENTS

INTRODUCTION

What Was an Environment?

Although there is a sense in which the environment has always surrounded us, these days it somehow manages to seem more ubiquitous than ever. On the largest of scales, climate scientists warn us of the consequences of greenhouse gas emissions for our global environment, while planetary scientists, aerospace entrepreneurs, and writers of speculative fiction imagine what would be required for us to flourish in the harsh environment of space or on the surface of other planets. Back on Earth, conservation biologists and the administrators of national parks and other protected areas manage the environments of endangered species and ecosystems across vast swaths of land and water, while people whose lives depend on forests and fisheries worry about the continental and oceanic environments that sustain those resources. In cities and other heavily settled areas, public health experts and card-carrying "environmentalists" seek to reduce air and water pollution and other ambient risks. Meanwhile, health and safety specialists provide guidance for creating the most supportive and productive environments possible within homes, schools, and workplaces. We even carry our own environments around with us: biologists claim that what the French physiologist Claude Bernard dubbed our *milieu intérieur* in the mid-nineteenth century is vital not only to the functioning of our own cells and organs but also to the survival of bacteria and other non-human members of our microbiome, which in turn help us regulate our relationship to our surroundings.[1] In that respect, we *are* environments just as much as we are *in* environments; we both surround and are surrounded.

If the environment seems to be waiting for us wherever we go, so does environmentalism, even if we limit ourselves to looking only for what one might call the "official" environmentalism of laws, regulations, treaties, government agencies, and nongovernmental organizations. Since the 1970s, environmental agencies and ministries have become fixtures of many national, regional, and municipal governments, while international environmental treaties have continued to grow in their coverage and complexity, even in the face of vigorous opposition. Beyond the legislative and diplomatic domains, new nongovernmental environmental organizations emerge on a regular basis, each competing with the others to inspire action, raise funds, and influence policies that will minimize or at least manage the harmful effects of human activities on the natural world and human health. In the private sector, advocates of corporate social responsibility argue that environmental and financial aims can be harmonized. In 2018, for example, the Starbucks coffee chain—the world's largest, with more than 27,000 stores—announced its Greener Stores initiative, which aimed to set "a new standard for green retail."[2] Meanwhile, in academia, environmental studies programs established decades ago are being revitalized alongside new programs in the environmental humanities.

Regardless of our involvement with or even support of official environmentalism, environmental concerns also shape the mundane details of our everyday lives. Many of us haul our recycling to the curbside in an attempt to compensate for a culture of disposability, or simply because we are legally required to, while the most motivated among us purchase carbon offsets to mitigate the impact of our air travel on the global climate. The design of the automobiles we drive—just like those of the buses, trains, and airplanes we ride—has been shaped by environmental legislation that seeks to conserve resources and minimize pollution through fuel efficiency standards and emissions tests. When we check into hotel rooms, we are offered the opportunity to help save the environment by reusing bath towels and declining housecleaning services, while in public restrooms we encounter paperless hand dryers and low-flow toilets accompanied by self-congratulatory environmental signage. The office buildings in which we work and the coffee we drink at our desks both come with green certifications. Those of us who make our living by hunting, fishing, or farming—or who seek out such activities in our leisure time—must often navigate thickets of environmental regulations before we can fire a gun, bait a hook, or plant a seed. To the extent that our identities are shaped by the material worlds we encounter on a daily basis, it seems we are all environmentalists now, whether we want to be or not.

At the same time, both the environment as a material reality and environmentalism as a social movement seem increasingly troubled. Even as some environmental problems are solved, the severity of others continues to mount. On the positive side of the balance, populations of some of the world's most iconic endangered species, including the bald eagle and the giant panda, are on the rebound, while rivers in many places are cleaner than they have been in decades. The use of certain kinds of toxic chemicals—including the bête noire of the environmental movement of the 1960s and '70s, the pesticide DDT— has been restricted to only the most urgent applications or eliminated entirely. The depletion of stratospheric ozone, which threatened to heighten human skin cancer rates by increasing the amount of ultraviolet radiation reaching Earth's surface, has been largely reversed. Take a step back, however, and such successes can seem like islands in a rising sea of environmental harms and haz- ards. Even as a few species recover, a few rivers grow cleaner, and a few toxic chemicals are banned, biodiversity continues to plummet globally, climate change accelerates, and microplastics, endocrine disruptors, and other new forms of life-threatening contamination of our shared surroundings continue to proliferate. As China, India, and other developing economies industrialize, moreover, some of the forms of air and water pollution that environmentalists in the developed world believed they had conquered decades ago are reemerg- ing on a vastly larger scale.

In part because of this mixed record, environmentalism as a social move- ment has also come under fire. Although it is difficult to find anyone, regard- less of where they stand on the political spectrum, who is opposed to a sound and healthy environment, there is little consensus over what that means or how to attain it. Surveying the state of environmentalism in 2008, the historian and activist Jenny Price described it as "a grab-bag of available causes and rhetorics old and new," including some she deemed to be of questionable value.[3] However inspiring environmentalism may have been in the early days of Rachel Carson and her fight against "biocides" such as DDT, Price and others have argued, the movement's apocalyptic imagination, precautionary pessimism, pervasive bureaucratization, and repeated failures to prioritize eq- uity and justice have weakened its ability to improve urban health, mitigate climate change, or respond to a range of other environmental threats. Even environmentalists of a less critical bent sometimes sink into pessimism, argu- ing that although the fight to save the environment was and is a noble one, it is time to admit that the battle has been lost and to begin adjusting to a new and diminished world. Meanwhile, critics from both the left and the right have characterized environmentalism (not without justification) as an effort to shift the burdens of development or the costs of quality of life from some people to others. The environment may be everywhere, but the question of how we should relate to it remains deeply contested.

QUESTIONING THE CONCEPTUAL FOUNDATIONS OF ENVIRONMENTALISM

Beneath these disputes and disillusionments lie even more fundamental doubts about the moral and conceptual foundations of environmentalism. One of these doubts concerns the very possibility and desirability of "saving the environment." Over the past several decades, an increasing number of scholars and activists have argued that, however well intentioned it may be, the impulse to "save" the environment reflects precisely the kind of hubris and sense of separation from the natural world that got us into trouble in the first place. Rather than treating the environment as if it were an object we can choose to ruin or save—that is, as something that is both separate from us and subject to our control—they argue that we should be learning to dwell responsibility within it. Since the 1990s, spurred both by the rise of right-wing and libertarian anti-environmentalism and by critiques from within, calls have grown louder to reorient the environmental movement away from saving pristine nature and toward taking responsibility for a world profoundly reshaped by human activity.[4] Some have even argued that the extent of that reshaping is so broad that the geological and historical epoch we live in ought to be called the Anthropocene, the age of humanity.

If the idea of saving the environment has been called into question even among committed environmentalists, so has the concept of environment itself. In fact, doubts about the value of the concept are not new. Since the emergence of the modern environmental movement in the 1960s and '70s, critics have argued that the concept encourages a spurious distinction between physical environmental problems such as pollution or resource exhaustion, on one hand, and social, economic, and political concerns, on the other. In the United States, for example, the home of one of the earliest and most vigorous national environmental movements, environmentalism was seen by some advocates of the antiwar, civil rights, feminist, and labor causes as a distraction. Among other things, they argued, it tended to gloss over the very real differences in the environmental challenges faced by different communities. By the 1970s, the anarcho-socialist theorist Murray Bookchin, who had embraced the concept of environment in his 1962 book *Our Synthetic Environment*, was encouraging his readers to focus instead on what he called "social ecology."[5] Compared to "environmentalism," he argued, "ecology" encouraged an approach to the human surroundings that was less instrumental and more attuned to matters of injustice and oppression.[6] Such critiques continue to be made today, leading some activists whose concerns might seem at first glance obviously environmental to eschew the term *environment* entirely.

In recent decades, even the viability of the concept of environment in a scientific context has been called into question. A growing number of biologists in particular have challenged the utility of dividing the world into organisms and environments and of seeking to explain the former in terms of their adaptations to the latter. Since the 1980s especially, biologists have developed a variety of metaphors, frameworks, and research programs that reject the conventional organism/environment distinction, working under labels such as developmental systems theory, niche construction, the Gaia hypothesis, and the extended evolutionary synthesis. The evolutionary biologist Richard Lewontin, for example, has argued that "genes, organisms, and environments are in reciprocal interaction with each other in such a way that each is both cause and effect," making it impossible to draw neat lines between them that are valid under all conditions.[7] Even if the concept of environment need not be entirely abandoned, he and other heterodox biologists have suggested, it needs to be radically rethought. Controversial when first introduced in the 1980s, such ideas have become increasingly mainstream in recent years.

Responding to these critiques from environmental activists and scientists as well as to developments within their own disciplines, scholars in the humanities have also questioned the value of thinking environmentally. In doing so, they have both built on and transcended a longstanding tradition of critiquing specific forms of environmental thought, from the intellectual historian and philosopher Georges Canguilhem's 1952 critique of mechanistic understandings of the "living and its milieu," to the anthropologist Tim Ingold's 1993 analysis of the paradoxes of the idea of "global environment," to the environmental historian Linda Nash's 2006 account of the emergence of a "modern" concept of environment as passive, homogenous, and clearly demarcated from the body.[8] At the heart of these critiques is a concern with the way the scientific concept of environment—which is also the concept of environment most often deployed in official environmentalism—seems to evacuate agency, experience, and embodiment from our understanding of life. Ingold, for example, has argued that conventional scientific understandings of the environment should be replaced with an embodied and local "mode of apprehension" that is "based on an active, perceptual engagement with components of the dwelt-in world, in the practical business of life, rather than on the detached, disinterested observation of a world apart."[9] Like the heterodox scientists mentioned above, these critics seek to rethink the concept of environment rather than rejecting it outright.

The fact that even many of the harshest critics of environmental thought have sought to somehow recuperate the concept reflects how deeply it has become embedded in our discourse. In recent years, however, the possibility of abandoning the concept entirely has been broached by a number of scholars.[10] If living beings are never really completely stable or self-contained, they suggest, it might be a mistake to place so much weight on a concept that divides the world into surroundings and the things they surround. "How on earth are you going to make the calculation of selfish interest and fit between 'an organism' and 'its environment,'" Bruno Latour asks, once you recognize "that the outside of any given entity (what used to be called its 'environment') is made of forces, actions, entities and ingredients that are flowing through the boundaries of the agent chosen as your departure point"?[11] Similarly, asks Donna Haraway, "what happens when the best biologies of the twenty-first century cannot do their job with bounded individuals plus contexts, when organisms plus environments, or genes plus whatever they need, no longer sustain the overflowing richness of biological knowledges, if they ever did?"[12] That such critiques are framed as questions suggests the difficulty of leaving the concept of environment behind; that they are being posed at all suggests that there are serious problems with the way the concept is being understood and used today.

HISTORICIZING THE CONCEPT OF ENVIRONMENT

Even if still tentative, these challenges to a concept that was long considered self-evident raise an important set of questions for historians who are concerned with past and present relationships between humans and their material surroundings. If the concept of environment does in fact profoundly misrepresent the nature of those relationships, how did it nonetheless become so central to the way we talk, think, and act? How far back in time, distant in space, or different in culture do we have to go to find people who have no use for it, and how close might be a future in which it is no longer of interest to anyone but historians? Most broadly, over the course of the concept's history, how have we changed not merely what we think *about* the environment but also what we think an environment *is*? These are questions to which historians concerned with changes in the material environment and in the ways humans have related to that environment—that is, "environmental historians," as they have been known since the 1970s—have paid surprisingly little attention. Even the enormous literature on the history of environmentalism, one of the central topics of environmental history, has barely touched on the history of the concept of environment, instead concentrating on the history of disputes over whether and how to protect an environment whose character and importance are assumed to be transparently obvious.[13]

This is not to say that environmental historians failed to critically examine any of their fundamental concepts. On the contrary, as the field expanded in the 1980s and '90s, they joined scholars in many other fields in the humanities and social sciences in questioning concepts they had hitherto taken for granted, from gender to technology to what was perhaps the master concept of the humanities and social sciences in the late twentieth century, "culture."[14] Paradoxically, however, environmental historians chose not to focus their critical attention on "environment"—the concept they had chosen for the name of their subfield—but rather on "nature," which to many of them seemed both synonymous with and more fundamental than "environment." Beginning in the early 1990s, the US environmental historian William Cronon led the field in challenging the idea that "nature," and particularly its embodiment in supposedly pristine "wilderness," was something that stood outside of human culture and could be used as a metric of human progress or a foundation of human history.[15] On the contrary, he argued, "nature" was a profoundly human concept with a history of its own. Over the course of the 1990s and 2000s, historians inspired by such arguments produced a number of "hybrid" environmental histories that started from the premise that nature and culture were always inextricably entangled.[16]

One of the curious consequences of the decision to focus on "nature" as a key concept was that "environment" almost entirely escaped examination in its own right. For those who assumed that the two terms were effectively synonymous and that environmental history could therefore be defined as the history of human relationships to nature, there was nothing surprising or concerning about this. On the contrary, by historicizing "nature," they believed that environmental historians had done the work necessary to allow their field to mature beyond its activist roots. Indeed, they argued, it gave them a critical stance from which to reevaluate the history of the environmental movement, and perhaps even to shape its future. If historical research showed that nature was always entangled with culture, then the environmental movement's focus on protecting only one form of that entanglement—that is, the kind of natural-cultural hybrid most visible in national parks and wilderness areas and other

places where evidence of human activity was at a minimum—was at best my-
opic and at worst actively harmful. A new and improved environmentalism,
they argued, would also attempt to protect the nature that was entangled with
culture in cities and suburbs, offices and factories, homes and neighborhoods,
and farms, forests, mines, and other working landscapes.

In addition to providing grounds for rethinking the contemporary environmental movement, environmental historians' focus on "nature" also shaped
their scholarship on the history of environmentalism. If environmentalism was
about an individual's or a society's relationship to nature, broadly conceived,
then its roots were both deep and broad. Not only could the environmentalism of the 1960s and '70s be seen as an extension and transformation of the
nature protection and conservation movements of the late nineteenth century
and early twentieth, it could also be seen as a continuation of much earlier
movements for the management or preservation of forests, water, wild animals,
or other aspects of the nonhuman world. Indeed, anywhere that historians
were able to find evidence that people had consciously attempted to ensure
that they were surrounded by conditions vital to their survival—that is, virtually everywhere in the historical record—they could claim to have found one
of the "roots" or "origins" of environmentalism.[17] Moreover, by focusing on
one or the other of these roots, historians could seek either to reinforce or to
challenge certain aspects of present-day environmentalism according to their
vision of how it could and should develop in the future.

It is only very recently that environmental historians have begun to turn
their attention to the history of the concept of environment as distinct from the
concept of nature. In doing so, they have begun to reveal a story that is quite
different from the ones they have told about environmentalism to date. In this
emerging story, environmentalism is not best understood as the modern man-
ifestation of a concern with nature that can be found in a diverse range of cultures but rather as something far more specific—namely, the practices, values,
and ideas that have coalesced among specific groups of people when they have
adopted the concept of environment as a foundation for understanding the
world around them. Often influenced by the history of science, this emerging body of scholarship assumes that objects of knowledge and concern such
as "the environment" have not always been conceptualized in the forms we
know them today but instead have emerged at particular historical moments
and have continued to change over time.[18] The aim of such scholarship is not
to add to the already enormous "roots and origins" literature, but rather to
explain how and why groups of people in various times and places have characterized their concerns in explicitly environmental terms and taken action

accordingly. The most ambitious and wide-ranging attempt along these lines is a collaborative project by Paul Warde, Libby Robin, and Sverker Sörlin that has resulted in a series of articles, a volume of primary sources with commentaries, and the book *The Environment: A History of the Idea*, which describes the emergence of "the environment" as the focus of scientific and political concern in the second half of the twentieth century.[19]

The present book builds both on the well-established body of scholarship that seeks the roots of environmentalism as commonly defined today and on this much smaller, more recent body of scholarship concerned with the historical emergence and transformation of the environment as an object of knowledge and concern. Like the former, it is motivated by a concern with today's urgent environmental problems, and it sees historical scholarship as one way of clarifying how we got here and where we are going. Like the latter, it relies on the historical record to find out what people in the past thought "environments" were and how those people protected, managed, improved, exploited, or otherwise interacted with them. However, while most of the latter body of scholarship focuses on the emergence of the notion of a singular, universal, or global environment in the decades following World War II—that is, "the environment" as we now usually conceive of it—this book begins its narrative in the late eighteenth century and includes a much wider range of variations on the concept of environment and the diverse environmentalisms that have been associated with them. The aim in doing so is to gain a better understanding of the past while also becoming more sensitive to the breadth of efforts underway today to reinvent the concept of environment for new needs and circumstances.

MATERIALIZING A MULTIPLICITY OF ENVIRONMENTS

Perhaps the most straightforward way to discover how and why people have adopted the concept of environment is to search for moments in history when they began to speak explicitly about "environments," to identify a particular set of concerns as "environmental," to describe themselves or others as "environmentalists," or to identify a theoretical framework or political ideology as "environmentalism." This word-centered approach quickly reveals that the concept of environment has a history that long predates the modern environmental movement but that is perhaps not quite as long as one might think. Although the word *environment* and its variants make occasional appearances in English texts as early as the beginning of the eighteenth century (and can be found much earlier in French), they did not come into wide usage in anything like their modern senses until the second half of the nineteenth century.[20] Indeed, *environment* sounded so awkward to early nineteenth-century ears that a friend of the Scottish writer Thomas Carlyle, who began using it idiosyncratically in some of his essays of the 1820s and 1830s, chided him for the appearance of what he considered to be a "positively barbarous" neologism.[21] By the late nineteenth century, however, many speakers of English found the term *environment* not only inoffensive but indispensable, even if the meanings they gave it and the stakes of the debates they had over it were quite different both from Carlyle's and from today's.[22]

Nor was this development unique to speakers of English. During roughly the same period, speakers of other European languages were adopting equivalent terms or beginning to use existing terms in similar ways, including *milieu* in French, *Umwelt* in German, and *ambiente* in Spanish. In other words, long before the emergence of the modern environmental movement in the mid-twentieth century, *environment* (or its equivalent in other languages) became a useful word for many groups of speakers. Following patterns of word usage can only take us so far, however. For one thing, it is obvious that people can share a concept even if they use different words to describe it, just as they can use the same word to express different ideas. As the various schools of the history of ideas, conceptual history, and intellectual history have taught us, while the appearance of new words can signal important conceptual shifts, determining the nature and significance of those shifts requires additional work.[23] In the case of the concept of environment, we need to remain open to moments when people are describing or encountering their surroundings in recognizably environmental terms even if they are not using the word *environment* itself.

This book takes two approaches to this problem. One is to work backward from people who explicitly used the term *environment* to others who, in the view of those historical actors themselves, had previously sought to express similar ideas in other terms. In the mid-1850s, for example, the British philosopher Herbert Spencer began using *environment* to describe the conditions to which individuals adapted. Although he denied the influence, it seems very likely that he borrowed the term from Harriet Martineau, who had first used it in *The Positive Philosophy of Auguste Comte*, her 1853 translation and condensation of the voluminous work of the French philosopher.[24] The French term that Martineau was translating was *milieu*, which Comte had begun using in a distinctive sense in the 1830s. Comte, in turn, based his understanding of *milieu* on decades of research by French naturalists into relations between what they called "organized bodies" on one hand, and their "conditions of existence" or "surrounding circumstances," on the other. Even though none of these earlier naturalists deployed the terms *milieu* or *environment* in the modern sense, they are an important part of the story of how those terms came into widespread use in the French and English languages.

This book also traces the concept of environment forward from people who explicitly used the term *environment* to others who were influenced by them. It does so by paying close attention to how the concept of environment has been embodied in practices, technologies, and social relations as well as in speech and text. In numerous cases, scientists have developed instruments and research practices on the basis of their understanding of the environment that have later been adopted by others who use them in similar ways for similar purposes even though they never deploy the term *environment* to describe what they are doing. Among the women who led the settlement movement in the United States at the end of the nineteenth century, for example, there were many who did not talk about their work in explicitly environmental terms but who borrowed techniques of social reform that Jane Addams and others had developed within an explicitly environmental framework. They are therefore an integral part of the history of the concept of environment. By following such tacit connections, this book offers what John Tresch has called a "materialized" intellectual history—that is, one that works from the premise that concepts become compelling and widely adopted because people put them into practice by transforming the material and social worlds around them.[25]

This is the very specific sense in which this book is a history of both environments and environmentalisms. That is, it is not mainly a history of how environmentalists have sought to protect the environment—a subject on which many books have been written—but rather a history of how the very idea of the environment has been materialized or put into practice in particular settings. In this sense, it is an environmental history of the concept of the environment, one that seeks to situate environmentalisms in various times and places. This may sound very similar to that staple of historical scholarship known as "contextualization," but there are some significant differences. Like contextualization, "environmentalization" helps us understand an otherwise isolated historical entity, event, or concept as part of a larger world or a longer or more complex narrative. Whereas the notion of context calls our attention to representation and interpretation, however, the notion of environment calls our attention to the material conditions that are essential for any entity, including a concept, to emerge and to persist. The context of environmental medicine in the British Empire consists of social, economic, political, and cultural factors; its environment is all of that, as well as climates, diseases, landscapes, technologies, and bodies—a list that future historians will likely find ways of extending or modifying, since just as there are many ways of conceptualizing the environment, there are many ways of environmentalizing the past.

If concepts come to matter only when they are materialized in particular environments, then what we mean when we use the word *environment* depends on the situation in which we find ourselves. That does not imply, however, that *environment* can mean anything we want it to. Even though there is an almost infinite diversity of ways to think and act environmentally, that diversity is constrained within certain limits. Language is flexible and changeable, but it is also the product of collectives of speakers who generally seek to remain comprehensible to one another. As a consequence, there are some patterns that hold true across the history of environmental thought, including the idea of a mutually constitutive relation between an entity and that which surrounds it—that is, a relationship in which each party not only influences the other but also in some fundamental way determines what the other is. In marked contrast to the term *nature*, which is often used to refer to the intrinsic character of a particular entity ("it is in its nature") or to aspects of the world that are fixed and unchangeable ("the order of nature")—that is, to things that are independent of any relation to external entities or forces—*environment* has almost always been used in this relational sense. This was expressed with particular clarity by the political scientist Lynton Caldwell, one of the architects of the first explicitly environmental legislation in the United States, who noted in 1963 that the "concept of environment assumes not only 'surrounding things' but something that is surrounded."[26] Without that fundamental relationality, the concept of environment loses much of its distinctiveness.

If the essentially relational nature of the concept of environment means that we cannot know what an environment is without knowing what it surrounds, it also means that as our understanding of the environment shifts, our understanding of what it means to be an entity surrounded by that particular kind of environment shifts along with it. We can see this necessary relationship between "surrounding things" and "something that is surrounded," in Caldwell's terms, at the very beginning of the history of the concept of environment, which required the invention of a new object of scientific inquiry: the "organism," which was defined as a combination of specialized parts ("organs") that worked together to allow a living being to survive and reproduce itself under a certain set of external conditions (its "environment"). We can also see this relationship in attempts emerging during roughly the same period to reconceive human populations as "a multiplicity of individuals who are and fundamentally and essentially only exist biologically bound to the materiality within which they live," as Michel Foucault wrote in describing new modes of governing subjects and citizens that emerged in the eighteenth century even before the concept of environment (or *milieu*) had been clearly articulated.[27] More broadly, each attempt to adapt the concept of environment to new circumstances and aims has been accompanied by changes in the understanding of the entities that are surrounded, whether those entities are imagined to be organisms, species, communities, civilizations, or the biosphere as a whole. The history of the concept of environment and the diverse environmentalisms associated with it is therefore also a history of the emergence of these kinds of surrounded entities, and of how various groups of people have imagined their ideal relationship to their surroundings.

THE SCOPE AND STRUCTURE OF THIS BOOK

Framed in such a broad way, this is a topic that could fill multiple books. Indeed, an account of all ways of conceiving of and relating to the environment would probably be as impossible to write as it would be to read. Fortunately, such a comprehensive account is not essential to conveying this book's core arguments—namely, that there have been many ways of being environmental since the emergence of the concept sometime between the late eighteenth century and the mid-nineteenth century; that particular ways of being environmental have emerged to serve particular aims under particular circumstances; that while none of these ways are either illegitimate or perfect, some of them are no longer very well suited to present-day aims and circumstances; and that we will as a consequence almost certainly need new ways of conceiving of and relating to our environments in the future, for which the past may serve as guide. For these purposes, a more modest selection of representative episodes suffices. There are links between each of the episodes, but this book does not present a narrative of smooth and continual progress, nor does it describe neat shifts from one paradigm to the next. Rather, it describes situated and partial adaptations and appropriations of techniques, practices, and ideas from one episode to the next.

A number of aims and constraints shaped my selection of episodes. One aim was to demonstrate that, far from being a universal concept with equivalents in every human culture and language and in all times and places, the concept of environment is the product of a very specific and—from a historian's perspective, at least—relatively recent history. I therefore decided that the book needed to begin in a period when no one was yet using the concept of "environment," or the term *environment*, in order to show that it was indeed possible to live and thrive as part of a community of speakers that had neither the word nor the concept. The book's first chapter therefore begins in the late eighteenth century, when the rough outlines of the concept of environment were first beginning to emerge, but when no word that carries the meanings of *environment* as we use it today (or even as it was used in the mid-nineteenth century) can be found in the written record. A corollary and perhaps at first glance contradictory aim was to show that both the concept of environment and various forms of environmentalism significantly predate the emergence of the modern environmental movement in the mid-twentieth century, and indeed significantly influenced those later developments. For that purpose, I tried to select episodes that demonstrate continuities from the earliest nineteenth-century uses of the term to the emergence of modern environmentalism more than a century later.

These aims were joined by practical constraints, including the need for access to historical documents in languages I could read and a body of scholarship that could help me interpret them. Reading texts in the original language was critical because translators often use familiar but anachronistic terms to convey what they believe the core meaning of a text to have been. There is nothing inherently wrong with this: from a contemporary reader's perspective, it may make little difference whether a nineteenth-century physician was writing about "external influences" or "the environment." For the historian of the concept of environment, however, the use of the latter rather than the former might indicate an important shift in perspective. Relying exclusively on translations can therefore produce a very muddled sense of how and when people began to speak, think, and act in explicitly environmental terms. As a result, this book focuses largely on speakers of European languages that I could read myself or was able to find aid in reading; most sources are therefore in English, French, or German, with a smattering of Norwegian and Russian. Another book (indeed, many books) could be written about the use of the concept by speakers of other languages, both European and non-European. For similar reasons, the book also focuses on historical actors and events that are extensively documented in the written record, which often reflect the dominant ways of being environmental in any given time and place. Yet another book could be written about the myriad alternative, minor, and quotidian environmentalisms that have flourished alongside and in opposition to these dominant forms.

Because the book ranges widely across time, place, and discipline—from the late eighteenth century to the present, from a museum in Paris to military stations in the British Caribbean, from social workers to UN diplomats, from natural history to climate science—each chapter must to some extent set the stage anew. To make both the connections and the differences between the episodes as clear as possible despite these leaps in time and place, each chapter follows more or less the same structure. After a brief introduction, there is a section describing the material and social conditions for a particular variation on the concept of environment and the form of environmentalism connected to it. The chapter on the development of the concept of *milieu* describes the importance of French imperialism for the growth of the collections of Paris's Muséum d'Histoire Naturelle (Museum of Natural History) in the 1790s, for example, while the chapter on the biosphere describes the impact of the resource crises of World War I on new theories of global flows of energy and materials in the 1920s. The following sections turn to the tools and practices—collecting, comparing, experimenting, mobilizing, modeling, and so forth—that materialized the environment for a particular community and gave them new ways of knowing and acting on their surroundings. Only after exploring these material conditions and situated practices does the chapter turn to the theories articulated by the people for whom the environment was a key concept—"environmentalists," as we might call them. The final section of each chapter considers how this particular set of tools, practices, and theories was later extended, transformed, or abandoned.

The book concludes by returning to the question with which we began—namely, whether the concept of environment can continue to serve us well, given the critiques that have emerged in recent years alongside growing concerns about the viability of the social movement; or whether it is time to set aside the idea that the world can be usefully understood in terms of entities, their surroundings, and the relationships between them. Having immersed myself for several years now in stories of people who adapted the concept of environment to circumstances and aims that earlier users of the concept could never have imagined, I tend to be optimistic about its future. It is undoubtedly true that as the world has changed, some of the most familiar ways of imagining and materializing the environment have become less compelling. That alone is not a reason to abandon the concept, however; as the two-century-long history of environments and environmentalisms shows, the concept has been refashioned over and over again to suit new aims and circumstances. Studying yesterday's environmentalisms makes it clear that there are many ways to be environmental beyond the ones we know today, and it may even give us some ideas for how to reinvent environmentalism for tomorrow. To begin the story in one of the many places where it could be begun, we now turn to Paris in the late eighteenth century, where the foundations were being laid for the development of the concepts of "environment" and "organism."

The World in the Museum: Natural History and the Invention of Organisms and Environments in Post-Revolutionary Paris

When the corvette *Géographe* docked at the French port of Lorient in the spring of 1804, it was at the end of an epic four-year voyage that had brought it from France to Australia and back, with stops in the Cape Colony, Mauritius, and Timor.[1] Carried out in the midst of a war between Britain and France that was roiling Europe and its colonies, the voyage of the *Géographe* and its companion ship, the *Naturaliste*, had not been easy. While the ships avoided armed conflict and in fact received a generally friendly welcome in British Australia, many of the crew members who embarked in 1800—including a number of naturalists tasked with surveying the landscapes, plants, animals, and peoples they encountered—fell seriously ill or died over the course of the expedition. They were "victims of their zeal," in the heroic gloss of one commentator; but also, more concretely, of dysentery, scurvy, and other misfortunes linked to exhaustion, malnourishment, unhealthy living conditions, and exposure to unfamiliar pathogens.[2] Even the expedition's captain, Nicolas Baudin, sickened on the return voyage and was buried in Mauritius.[3]

For those who had eagerly awaited the return of the *Géographe* and the *Naturaliste*, the expedition's spectacular results fully justified its high cost in blood and treasure.[4] Among them were the naturalists of Europe's preeminent natural history museum, the Muséum National d'Histoire Naturelle in Paris, who had been closely involved in the planning of the expedition.[5] In addition to a detailed map of the coast of Australia that promised to aid France in its global struggle against the British, the expedition brought back sketches and descriptions by its team of naturalists, as well as dozens of living animals, hundreds of plants, and case upon case of seeds, dried plants, rocks, bones, and other natural artifacts.[6] "Of all the collections that have come to us from distant lands, at diverse epochs," the botanist Antoine Laurent de Jussieu wrote as the cataloging of the collection was beginning, "the one brought by the vessels the *Naturaliste* and the *Géographe* is certainly the most substantial."[7] The museum's botanists, zoologists, geologists, and other naturalists would spend years working through it.[8]

Although the number of new plant and animal species the Baudin expedition introduced to France was unusually high, as Jussieu's comment suggests, the expedition itself was not unique. Over the course of the eighteenth and nineteenth centuries, European museums, menageries, and botanical gardens swelled with specimens gathered in the course of encounters between explorers, merchants, soldiers, and colonists and new places and peoples. Six years before the *Géographe*'s arrival in Lorient, for example, Baudin himself had brought the museum a collection of natural history specimens from the West Indies that included a number of living trees, some of them ten feet tall.[9] Such large acquisitions stood out against a background of smaller but more regular contributions from a global network of correspondents and collectors. Meanwhile, the institutions that housed these collections were generously supported by monarchs and republican governments alike, for reasons that were clear to all. These institutions symbolized the advancement of knowledge, the power of empires, and the global reach of trade networks while also promising practical economic benefits, such as the development of new kinds of crops and domesticated animals.

By the end of the eighteenth century, the museum and other institutions like it had become central to answering basic questions about the organizing principles of nature and to developing new tools and methods for reshaping nature's order in the service of human desires. They also served as sites for the emergence of a new understanding of life that centered on the interaction between organized bodies and the influences that surrounded and shaped them—an understanding that would eventually lead naturalists to develop a new concept, "environment" (or in French, *milieu*). While naturalists had long recognized that the character of organisms could be seen as related to "climate" and "place"—understood as the atmospheric conditions prevailing across each region of Earth and as the combination of weather, plants, animals, and people that made up the character of a given location, respectively—the concept of environment was something new. More than just the physical surroundings, it referred to the aspects of those surroundings that were significant for a particular "organism" (or *organisme*)—another novel concept that gradually emerged during this period to become the focus of naturalists' attention.

FIGURE 1. The *Géographe* and the *Naturaliste* anchored at Kupang in Timor in 1801. (Reprinted from plate 39 in Charles-Alexandre Lesueur and Nicolas-Martin Petit, *Voyage de découvertes aux terres australes* (Atlas historique) [Paris: Imprimerie Impériale, 1807].)

Far from being self-evident, the idea that life could be described in terms of the relationships between "organisms" and their "environments" was the product of a diverse but connected set of observations and experiments at a particular moment in history. It depended on new techniques of collecting, classifying, and experimenting on living things at institutions such as the Muséum d'Histoire Naturelle, whose scientific discoveries, research methods, and novel philosophical frameworks for understanding life provided models that others would later adapt to their own ends and circumstances. By examining both the material practices and conceptual frameworks that these naturalists developed between the late eighteenth century and the mid-nineteenth century, we can see how the concept of environment came to matter to a community of people who had previously made do without it, and how it differed for them from other concepts such as climate and place that at first glance might seem similar. Knowing that the concept of environment was invented and painstakingly put into practice at a particular time and place helps us see that it is neither universal nor timeless, and that it is not our only option for understanding and engaging with our surroundings today.

COLLECTING

Natural history at the end of the eighteenth century was far from the tedious, stagnant, and amateurish enterprise of later caricatures. On the contrary, it was a highly dynamic enterprise that linked museums, botanical gardens, and other institutions to global networks of correspondence, exploration, conquest, and trade.[10] Naturalists at the Muséum d'Histoire Naturelle and similar institutions elsewhere were responsible for assembling and classifying diverse collections of plants, animals, minerals, fossils, and other productions of nature that Europeans encountered. By gaining insight into the basic order of nature, they hoped to be able to reshape it in ways that benefited themselves and the states that patronized them. At a time when European economies were based on agriculture, expertise in the identification, growing, and breeding of plants and animals was anything but impractical. On the contrary, it went to the heart of the preindustrial economy. As specimens from around the world flooded the museum's anatomical galleries, herbaria, and seedling gardens, naturalists had both opportunity and motivation to devise new methods of organizing and explaining nature's diversity. These new methods would eventually result in the emergence of "environment" as a central concept for the study of life.

The Paris museum played a leading role in the development of new methods and theoretical frameworks for understanding the conditions that enabled organized bodies of particular kinds to survive and flourish. Officially founded in 1793, the roots of the Muséum d'Histoire Naturelle lay a century and a half earlier in the establishment of a royal garden for medicinal herbs south of the Seine River, which was later complemented by a collection of natural history specimens kept in a building known as the Cabinet du Roi.[11] From these relatively modest beginnings, the garden and the natural history collection were transformed into a highly productive research center from the 1730s onward under the leadership of Georges-Louis Leclerc, Comte de Buffon, a naturalist whose fame in the eighteenth century was rivaled only by that of Linnaeus.[12] Drawing on the far-flung networks of trade and conquest that were filling the streets and shops of Paris with exotic animals, plants, and products, Buffon and other mid-eighteenth-century French naturalists capitalized on the expanding reach of European empires to gain unprecedented access to the world's biological diversity.[13]

Even amid the political turmoil of the years before and after the Revolution of 1789, when any institution with royalist associations was at risk, the reputation of the garden and the natural history collection as unparalleled sources of natural knowledge helped ensure their survival. Indeed, the museum that was established in 1793 on the foundation of the garden, the natural history collection, and a new menagerie of living animals flourished as an icon of the Enlightenment values of the French Republic. In this way, the garden and natural history collection emerged from the Revolution even more prestigious and more thoroughly entwined with the aims of the French state than before.[14] In the years following the museum's establishment, its naturalists became increasingly implicated in France's imperial ambitions. During Napoleon's ill-fated expedition to Egypt in 1798, for example, naturalists such as Étienne Geoffroy Saint-Hilaire gathered specimens and made observations that significantly enriched the museum's collection.[15] In this regard, the Baudin expedition—which aimed both to map the terrain of future naval conflicts and to gather specimens that would reveal the order of nature—was typical of the French natural history of its day.

France's imperial ambitions also aided the growth of the museum's collections through the acquisition of the collections of other empires, as France annexed or invaded other European states over the course of the Napoleonic Wars. The incorporation of items from the Dutch stadtholder's collection into the museum after 1795, for example, gave French naturalists control over one of the richest natural history collections in Europe, including specimens from Dutch colonies and trading ports in Indonesia and from other places beyond the bounds of the French empire.[16] In effect, French naturalists were able to fill in the gaps in their own collections by exploiting the networks established by other empires. One scholar has estimated that the specimens acquired from Dutch, Spanish, Portuguese, Belgian, and Italian institutions constituted more than a fifth of the museum's new acquisitions between 1793 and 1809.[17] The museum's naturalists were not shy about celebrating their gains, arguing that they gave them a view of nature's diversity unrivaled at any other European institution. In 1795, for example, André Thouin, professor of horticulture as well as head gardener at the museum, celebrated the acquisition of the Dutch specimens, which he argued would make the museum's collection "the most magnificent in the world and the most useful for the progress of the natural sciences."[18] These specimens included the skeletons of a rhinoceros, a giraffe, several kinds of monkeys, and numerous other animals.[19]

The museum's naturalists also benefited from the natural history research of other nations even when they did not acquire foreign collections by force. Both before and after the Revolution, the museum's naturalists enrolled a wide network of paid correspondents eager to benefit from their expertise and contribute to their research. First initiated under Buffon's intendancy, this network expanded over time, with the museum issuing increasingly detailed and specific guidelines for its correspondents and collectors to follow. For example, an 1818 pamphlet, *Instructions for Travelers and Officials in the Colonies, on the Manner of Collecting, Conserving, and Shipping Natural History Objects*, provided guidance to ship captains and colonial officials on "obtain[ing] in the diverse countries where they will sojourn the objects needed by the Muséum."[20] Birds, for example, needed to be killed with the proper size of shot to avoid damaging the specimen; the blood then needed to be drained, a piece of cotton placed in the beak, and the bird skinned as quickly as possible to prevent the feathers from detaching during putrefaction.[21]

Although it was the museum that collected the specimens and set the standards, the relationship between Parisian naturalists and their collectors and correspondents elsewhere was often reciprocal and mutually beneficial. The correspondents enriched the environment of the museum's naturalists with their specimens and reports of distant climes, meanwhile gaining access to the new techniques and ways of understandings of the natural world that were being developed at the museum. As a result, many colonial officials and foreign naturalists contributed to the museum's store of knowledge and willingly submitted to its standards without any expectation of monetary remuneration. In the first decade of the nineteenth century, for example, the museum could count among its correspondents the sitting president of the United States, Thomas Jefferson, who had become acquainted with Buffon in Paris in the 1780s and continued to correspond with French naturalists after his return to the United States.[22] Through the publication of Jefferson's reports on the development of a new type of plow or on Lewis and Clark's expedition to western North America in the museum's journal, French naturalists were introduced to organisms, places, and practices that they would likely never encounter personally.[23] At the same time, Jefferson's connection to the museum helped him stay abreast of the most advanced theories and findings in natural history and gave him opportunities to assert the value of the North American continent and the United States' interest in it, including the vast territories acquired from France as part of the Louisiana Purchase in 1803.[24]

With their long experience in diverse places around the world, settlers and colonists such as Jefferson were invaluable sources of specimens and reports for the museum naturalists, who were only too willing to ignore—or even celebrate—the violent and exploitative relationships that made much of that knowledge possible. In the course of managing slave plantations, coordinating the extraction of natural resources, or surveying land for potential settlement, agents of Europe's imperial expansion were able to observe the responses of plants, animals, and human health to subtle shifts in climate and other conditions across periods of years or even decades. This distinguished them from the museum's traveling naturalists, who had few opportunities to observe how those places or the plants, animals, and people living in them changed over time, and were therefore forced to rely on the knowledge of locals even when they were able to visit those sites themselves. During a five-month stop in Australia, for example, naturalists with the Baudin expedition turned to British colonists to learn about the surrounding landscape and its flora and fauna, often bartering for specimens and soliciting information about them secondhand rather than carrying out their own collections and observations in the field.[25]

Much of the information that the museum's naturalists were able to acquire from European settlers was a hybrid of the latter's firsthand experiences and the knowledge they had gleaned from people who labored or lived on that land. In Port Jackson, for example—today the site of Sydney—British settlers lived in close proximity to Aboriginal Australians and occasionally adopted their names for local species. The scientific name of the species now known as the southern elephant seal, *Mirounga leonine*, for example, is based on the indigenous word *miourong*.[26] In that sense, knowledge produced at the museum was sometimes indirectly derived from non-Western traditions of indigenous knowledge. The museum's traveling naturalists also turned directly to indigenous inhabitants of the places they visited for insight, establishing relationships that were highly asymmetrical but sometimes mutually beneficial. While naturalists gained knowledge and specimens, indigenous collaborators stood to gain money, goods, status, and new allies. In Timor, for example, François Péron, the lead naturalist of the Baudin expedition, learned about marine life and acquired specimens from a local fisherman named Néâs.[27] At the same time, such local informants were often objectified and condescended to by the very same naturalists who turned to them for help; even as Péron sought out the advice of Néâs and other local informants, he also subjected them to anatomical and ethnographic studies, thus transforming them from collaborators into objects of research.[28]

At the end of the eighteenth century, the museum's access to specimens and descriptions of living beings and the conditions of their lives from around the world was unparalleled. Dependent on the spread of European empires, its collection of specimens and the resulting knowledge relied in large part on the hidden labor and knowledge systems of the non-European peoples who lived and labored on the lands that European colonists had settled. In these ways, the rise of institutions such as the Muséum d'Histoire Naturelle reflected both the opportunities that European imperial expansion had created and the economic and military needs of European empires. No matter how abstract or theoretical the debates among naturalists or how universal their ambitions to survey the variety of living beings in existence might have been, ultimately their work was built on a foundation of imperial trade, settlement, agriculture, and military conflict. Under these conditions, it is not surprising that natural history collections were centralized at institutions such as the museum, or that naturalists at such institutions would increasingly seek to identify the essential conditions that allowed living beings of particular kinds to thrive far from the places where they were originally found.

CLASSIFYING

When the *Géographe*'s cargo of preserved specimens and living plants and animals was delivered to Paris in the spring of 1804, one challenging task had just been completed, but another equally imposing one remained: unpacking, identifying, describing, and classifying the new acquisitions.[29] The Baudin expedition had included its own team of voyaging naturalists, but properly describing and classifying their discoveries was something that French naturalists believed could be accomplished only at the Muséum d'Histoire Naturelle. Indeed, classification was the central work of institutions such as the museum during this period, offering both a conceptual framework for making sense of nature's order and a practical tool for organizing and maintaining botanical gardens, menageries, and museums. Both abstract theories of taxonomy and everyday practices of organizing specimens for research and display allowed naturalists to recognize essential similarities and differences across species and to relate those variations to the conditions under which species were found in nature. Over time, the introduction of new systems of classification therefore helped refashion eighteenth-century institutions of natural history into sites for studying the relationship between the structure and function of living beings and the conditions that surrounded them.

In the mid-eighteenth century, the anatomical specimens in the Cabinet du Roi had been arranged primarily according to two criteria. The first was general similarity, which was judged according to the sum total of all their traits; the second was their familiarity and utility in French society. Trees were considered alongside other trees rather than with smaller plants, even when the latter had similar flowers or other characteristics, and the familiar horse and domestic cat were considered separately from the exotic zebra and lion, despite their obvious anatomical similarities. This was roughly the approach to classification that Buffon had advocated in the first volume of his *Natural History, General and Particular*, which appeared in 1749 and was followed by a series of widely read volumes over the succeeding decades. It stood in stark opposition to the kinds of formal classification schemes based on a few key traits that had been proposed by Linnaeus and others in the mid-eighteenth-century— so-called "artificial" systems that Buffon criticized for seeking clarity and simplicity at the cost of truth to nature.[30]

After Buffon's death in 1788 and the establishment of the Muséum d'Histoire Naturelle in 1793, French naturalists reconsidered both their schemes of classification and the arrangement of the museum's specimen collection and gardens. As the number of new specimens grew (along with the demand for access to them for purposes of research and teaching), the expansion and reorganization of the galleries and garden became increasingly imperative. Under Buffon's intendancy, the greenhouses and other facilities had been significantly expanded, but by the 1790s they were showing their age after years of neglect during and after the Revolution.[31] Moreover, many of the older specimens in the anatomical collection were decaying as a result of poor preservation techniques, even as they occupied space that was needed for new acquisitions.[32] In 1798, Jussieu, the museum's director, was concerned enough to warn the French government that the museum was "menaced with imminent destruction."[33] In the late 1790s, the museum extended its gardens onto newly purchased land and renovated old buildings on an adjacent lot once used for the administration of Paris's hackney coaches to provide larger galleries for the expanding collection of anatomical specimens.[34] These efforts continued after Napoleon's final defeat in 1815 and the subsequent restoration of the Bourbon monarchy, when the space devoted to the anatomical galleries was expanded even further.[35]

Beyond simply increasing the amount of space available for the gardens and the anatomical collection, the museum's naturalists also fundamentally reconsidered the classification scheme that they used to organize their collections of preserved and living specimens. At the instigation of Georges Cuvier, who began working at the museum in 1795, the anatomy collection was transformed so that the functional organization of each of the species represented in the collection—that is, the relationships among its "organs"—became the primary criteria for classification and arrangement, replacing the criteria of similarity and familiarity that had been used during Buffon's time. Cuvier's approach emphasized biological functions such as digestion, locomotion, or reproduction as part of "comparative anatomy," a new research program that sought to identify distinct patterns of organization across the diversity of living beings. In 1803 he announced that the museum's anatomical specimens were now being "arranged with physiological aims, that is, they were divided not primarily according to the class of animals from which they arose, but according to the organs whose structure they elucidated."[36] Indeed, it was mainly the desire to make such functional relationships visible, rather than simply the need for additional storage and display space, that drove the expansion of the anatomical galleries from the 1790s onward.[37]

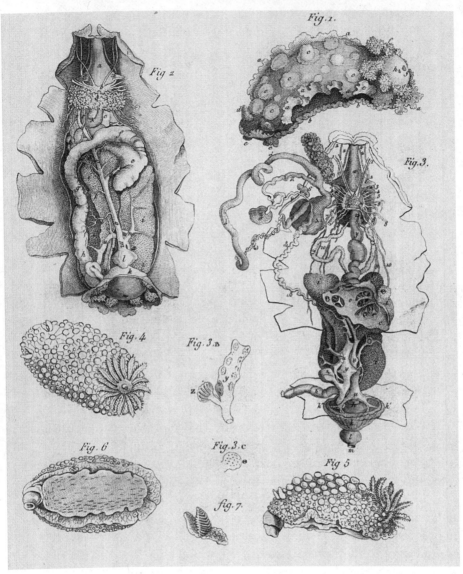

FIGURE 2. Illustration of a dissection by Georges Cuvier of a sea slug in the genus *Doris*, showing the size, shape, and relative position of various internal organs. (Reprinted from plate 73 in Georges Cuvier, "Mémoire sur le Genre Doris," *Annales du Museum National d'Histoire Naturelle* 4 [1804]: 447–73.)

The reorganization of the specimens in the new comparative anatomy galleries was linked to a shift in the kind of anatomical research conducted at the museum. Most research continued to consist of the patient work of carefully preserving, dissecting, and describing the forms of once-living beings. Naturalists used such specimens and descriptions to compare the shape and configuration of the organs of living beings of different species and to elucidate the fundamental principles governing their structure and development. Increasingly, however, the focus of research was not simply on the superficial appearance of an organism's parts but also on the physiological functions performed by those parts. The research of one of Cuvier's students, Henri Marie Ducrotay de Blainville, is particularly illustrative. A beneficiary of the museum's period of dramatic expansion in the early nineteenth century, Blainville began his career in the 1810s as a protégé of Cuvier and went on to become one of France's most influential zoologists. Over the course of his career, Blainville's careful dissections and descriptions of animals increasingly focused on the functions performed by each organism's organs rather than simply on their comparative size, shape, and texture.[38]

Research on the living specimens kept in the garden and the menagerie also shifted in response to new methods of classification centered on the functional organization of living beings. Head gardener Thouin, for example, organized his plantings not according to their appearance or place of origin but rather according to the external conditions that each kind of plant required to thrive.[39] Conventional distinctions between climatic zones, such as "tropical" versus "temperate," still provided a rough scheme of organization, with plants from warmer climates positioned to maximize their exposure to the sun, but other factors also came into play in the arrangement of the gardens, including temperature, water, wind, sunlight, and soil type. Ultimately, what mattered to Thouin was not where a plant had been found but what conditions he and his fellow gardeners needed to provide for it to survive. Plants from widely varying climates could be grown in the same or adjacent plots, as long as the set of conditions present in those plots was compatible with their growth.

The naturalists in charge of the museum's collection of living animals shared many of the interests and aims of those working in the gardens and anatomical galleries, but faced even more daunting challenges in pursuing them. One challenge arose from the haphazard nature of the collection. In the first decade of the menagerie's existence as a formal part of the museum, there had been very little order to it. Some of its animals were acquired when the practice of displaying exotic animals for private profit on the streets of Paris was banned following the Revolution, while others were drawn from the former royal me-

nagerie at Versailles or from expeditions such as Baudin's.[40] Consequently,
the naturalists in charge of the menagerie had even less control over the range
of species in their living collection than Thouin and his fellow gardeners did
in theirs. Moreover, even in those rare cases where the menagerie held multiple specimens of a given species, the cramped conditions contributed to high
mortality rates and reduced the chances of reproduction. In comparison to the
garden's abundance of plant life, opportunities for classification and experimentation in the menagerie were therefore quite limited.

Nonetheless, to the extent possible, the menagerie keepers sought to organize their living collection along principles similar to those of the galleries
of comparative anatomy—that is, in a way that emphasized not similarities
in the external appearance of different species, but rather the various ways in
which they accomplished the functions necessary to life. This was particularly true after Georges Cuvier's younger brother Frédéric was appointed to
the position of *garde* (keeper) of the menagerie in 1803.[41] Under his watch,
the menagerie served as a resource for the museum's comparative anatomists,
whose observations of its living specimens and dissections of those who had
died gave them insight into the anatomy of fossil animals and preserved specimens. After a leopard died in 1794, for example, its body was dissected and
sketches of its internal organs were made by an artist working for the museum.[42] Later, in preparation for a study of the fossil remains of large cats,
Georges Cuvier studied both preserved specimens in the collections and a
living panther that had been collected in Java.[43]

The concern with functional organization as the basis of the classification of
plants and animals also shaped the way naturalists presented their findings vi-
sually in the richly illustrated journal *Annales du Muséum National d'Histoire
Naturelle*. By eliminating incidental detail relevant only to a particular specimen and by minimizing references to landscapes, historical settings, cultural
artifacts, and exotic peoples, naturalist-illustrators of the late eighteenth century and early nineteenth century aimed to make the most essential characteristics of a plant or animal as visible as possible, even if that meant presenting an
ideal type rather than a specific specimen.[44] In doing so, they sought to bring
the inner logic that governed the structure of each type of organism into sharp
focus. That, in turn, raised the question of why each organism was organized
the way it was, a question that many of the museum's naturalists concluded
could be answered only by relating its anatomical structure and function to the
external conditions under which it lived. Paradoxically, then, the exclusion of
contextual detail from illustrations of plants and animals provoked naturalists

to think in new ways about those organisms' conditions of existence, both internal and external.

By examining the methods of classification that naturalists developed at the museum to manage the flood of specimens that was pouring into their collections and to pursue the research questions they found important, we can see how practical concerns were intertwined with theoretical questions and with the processes driving French imperial expansion during the Napoleonic years. The advantages and disadvantages of various systems of classification were debated in lecture halls and in the pages of scholarly journals, but they were also hashed out through decisions to acquire land, dig ditches, construct buildings, build cabinets, and shuffle living and preserved specimens from place to place. Such material interventions and everyday practices helped make plausible the division of the natural world into two parts—one consisting of living beings ("organisms") and the other consisting of the conditions and surrounded and shaped them ("environments")—even before naturalists had the words to express what they were doing.

EXPERIMENTING

From the late eighteenth century onward, interest among naturalists at the Muséum d'Histoire Naturelle regarding the function of each of an organism's distinct parts (its "organs") and the organization of its body as a whole led them toward experimentation as a method of research. By testing the effect of various external influences on the development or health of a plant or animal, naturalists hoped to gain greater insight into why it was organized as it was. At the same time, experimentation was a practical necessity in the garden and the menagerie, where plants and animals had to be kept alive under conditions that in most cases differed dramatically from those under which they had been found. Moreover, because agriculture was the foundation of the French economy during this period, such experiments had potentially enormous economic and strategic implications. If exotic spices, fruits, vegetables, or medicinal plants could be grown in metropolitan France or one of its tropical colonies, for example, the shape of France's trade networks and the global balance of power might be altered. Experimentation was thus a path toward greater insight into nature's order, a prerequisite for research on certain living plants and animals, and a potentially revolutionary intervention into the economy of the French Empire.

In a sense, experimentation with the relationship between organisms and their environments began even before specimens arrived at the garden or the menagerie, inasmuch as it was required to keep plants and animals alive on their journeys to Paris. Although any attempt to translocate a living being carried risks, multiyear collecting expeditions such as Baudin's faced especially daunting challenges. Over extended periods, ships such as the *Naturaliste* and the *Géographe* became floating ecosystems within which only certain kinds of plants and animals survived, and which indeed often challenged the health of their human occupants.[45] The practical challenges inherent to keeping plants and animals on a seagoing ship were often compounded by ignorance of the conditions required for each kind of organism to survive. Furthermore, even when precise descriptions of the conditions under which organisms had been originally found were available, those conditions were usually impossible to replicate for practical reasons, regardless of how well-equipped, knowledgeable, and conscientious the collector or the crew might be.

As a result, many valuable specimens perished long before they reached the garden or the menagerie. The *Naturaliste*, for example, was loaded with specimens and sent ahead of the *Géographe* to France in 1803 while the latter completed its surveying and collecting mission, but many of the expedition's painstakingly collected living plants withered before they could be seen by the museum's naturalists. As the ship approached France, it was accosted by a British ship that forced it to a port in England, where it lost "precious time and many living plants."[46] Only twenty of the eight hundred plants originally on board reached Le Havre alive.[47] The *Géographe*'s living collection fared better, but even it suffered significant losses. On the ship's arrival in France, contrary winds forced the captain to dock at Lorient rather than sailing up the Loire, necessitating a lengthy overland voyage to Paris. Despite being personally accompanied by Geoffroy, many of the plants and animals that had survived their time at sea died before they reached the museum, while others lasted only days or weeks after their arrival in Paris.[48]

Indeed, once seeds or living specimens had arrived safely at the museum, the challenge of keeping them alive continued, albeit under easier circumstances. In order to determine the conditions that would allow them to flourish far from their places of origin, naturalists subjected them to a variety of experiments. Among the sites at the museum for carrying out such experiments, the garden's seedling plots and hothouses were the most important. Established by Buffon in 1785, just before the Revolution, the seedling garden covered more than 3,300 square meters and was divided into numerous smaller plots devoted to the raising of plants requiring different kinds of conditions to survive.[49] Given the diversity of species in the collection, ensuring the survival of

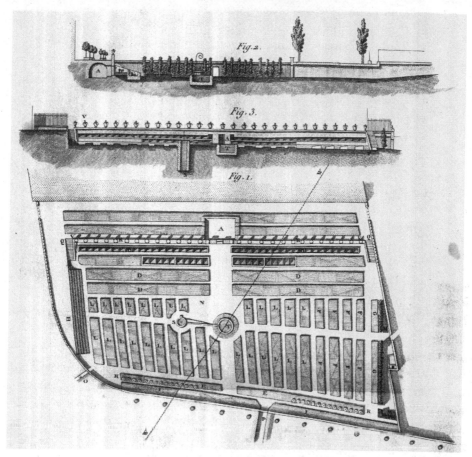

FIGURE 3. Vertical cross-sections and a bird's-eye view of the seedling garden at the Muséum d'Histoire Naturelle, circa 1804. (Reprinted from plate 62 in André Thouin, "Description du Jardin des Semis du Muséum d'Histoire naturelle, de sa culture et de ses usages, première partie," *Annales du Muséum d'Histoire Naturelle* 4 [1804]: 263–88, description on 287–88.)

seedlings and encouraging their acclimatization to Parisian conditions were constant concerns for Thouin, the head gardener. Among all the garden activities, the raising of plants from seeds was, he wrote in 1805, "the most complicated, because it demands a great number of processes as varied as those of nature, for all the series of plants of all the parts of the globe."[50]

The effort to keep all these plants alive, including trial-and-error experimentation with soil, moisture, shade, and other factors, gradually led Thouin and other naturalists working in the garden to an understanding of environment as distinct from both climate and place—that is, it led them to an interest in the relationship between a living being and the conditions relevant to its survival. Although climate provided a general scheme of organization for the seedling garden, Thouin recognized that replicating all the world's climatic zones within the confines of the garden was neither feasible nor sufficient for the growth and reproduction of plants. Instead, he argued, "one must limit oneself to imitating the essential properties of the principal climates."[51] Creating the proper set of conditions for each type of plant required a range of techniques, many of them borrowed from practical gardeners.[52] For example, Thouin and his staff found that certain dark-brown soils, which were particularly effective at absorbing and retaining heat, were appropriate not only for tropical plants but also for temperate-zone plants whose roots were deep enough to reach the cool soil below.[53] They also developed a variety of designs for what they called *contresols*—shaded containers that allowed a potted plant to receive precisely the amount of wind, rain, and sunlight it needed to thrive.[54] Such techniques embodied each plant's environment as a set of impinging external factors that could be reproduced on the garden's grounds in Paris, rather than as climatic zones or geographical locations on the surface of the planet.

Despite the skill of Thouin and his staff and the resources to which they had access, however, there were limits to their ability to keep exotic plants alive in Paris. Even hothouses warmed at great cost during the Parisian winters could not provide a hospitable environment for all the world's plant species or simulate the growth of any given species under all possible climatic and soil conditions. Aquatic plants, for example, proved difficult to raise, as did fungi; indeed, the latter were so difficult to raise that Thouin and his colleagues resorted to creating models out of lead for the purpose of instructing students.[55] To compensate for these limitations, the museum established connections to a global network of botanical gardens. In 1807, one naturalist reported that the museum had been in contact with ninety-seven botanical gardens during the previous two years—fifty of them within metropolitan France, and the remaining forty-seven elsewhere in Europe and as far afield as New York, Calcutta, Cayenne, and Mauritius.[56] The result of exchanges among these botanical gardens, the naturalist predicted, would soon be the "naturalizing in all civilized countries those among the useful plants, the culturing of which is not invincibly opposed by the difference of climate."[57] Crucial to this program of acclimatization was the fact that the traffic of specimens between this and other gardens was bidirectional, each providing to the others plants that grew best in the climate and conditions where they were located. In 1808, for example, the director of the botanical garden at Cayenne in French Guiana wrote to Thouin to report on the success of his effort to acclimatize the breadfruit trees that he had received from the Paris museum a decade earlier.[58]

The menagerie also served as an important site of experimentation, although in comparison to the garden its possibilities were highly constrained by the relative difficulty of transporting and caring for live animals. As with plants, however, a certain amount of experimentation was necessary simply to keep the animals alive once they had reached Paris. Those involved in maintaining the collection often resorted to trial and error or drew on their experiences with domesticated animals to do so—domesticated donkeys and dogs serving as analogues for African zebras and Australian dingoes, for example.[59] In the process, these naturalists came to realize that their task was to identify the conditions most essential to an organism's survival rather than trying to recreate its native climate or place of origin in its entirety. Even though it was clearly impossible to reproduce the African savanna or the Australian outback in the heart of Paris, they believed it was feasible to reproduce the essential conditions that would allow species from those climates and places to thrive.

Menagerie naturalists also experimented with the possibility of adapting animals to the Parisian environment by interbreeding them with domesticated species in ways that would alter their internal organization just enough to allow them to survive under altered conditions. Just as the Parisian environment could be refashioned to support the flourishing of exotic organisms, they argued, so could those organisms be refashioned to suit the conditions of Paris. As the keeper of the menagerie, Frédéric Cuvier supervised a number of such attempts at hybridization, which he argued were the key to acclimatizing animals to new climates. Among these experiments were an attempt to mate a dingo with a dog, a zebra with both a donkey and a horse, a whooper swan with a domesticated goose, and a wolf with a dog.[60] Such experiments, though rarely successful, were indicators of naturalists' faith that the relationship between an organism and the conditions in which it thrived was potentially malleable given the proper techniques.

Although the museum's comparative anatomists focused their attention on bones, skins, and other preserved specimens—that is, on dead rather than living animals—they too were attracted by the possibilities of experimentation, which promised to shed light on the functional organization of living beings. Such findings might, they believed, have important implications for the classification of fossil remains and preserved specimens. In the late 1790s, for example, while stationed in Cairo with the French expeditionary force, Geoffroy either conducted or planned several experiments on the effect of external conditions on an organism's form and function. These included one experiment in which Geoffroy sought to influence the number and distribution of pistils and stamens in a plant's flowers by modifying the amount of light and space it received.[61] Such experiments were impossible to carry out with preserved specimens, but they nonetheless shaped the way Geoffroy and others interpreted the variations in form that they observed in the comparative anatomy collection.

By examining the experimental techniques that naturalists developed to keep plants and animals alive in transport and to raise them successfully in the garden and the menagerie, as well as the kinds of targeted experiments that Geoffroy, Frédéric Cuvier, Thouin, and others pursued to determine the limits of plant and animal adaptability, we can see how a new kind of scientific object—or, rather, a new pair of scientific objects—emerged in a particular time and place. The concepts of "organism" and "environment" were not simply new ideas applied to the world as it had always existed; rather, they were descriptions of a world that was in the process of being changed by new practices, techniques, and settings. French naturalists' efforts to transplant living beings to Paris from around the world were dependent on the global networks of imperial conquest and trade that had expanded over the previous centuries, as well as on the techniques they developed at the museum from the end of the eighteenth century onward. "Environment" was an idea crafted precisely to describe this emerging world of mobile organisms and experimental environments.

ORGANISMS AND ENVIRONMENTS

The naturalists who sorted through the *Géographe*'s and *Naturaliste*'s natural history specimens in the years after 1804 did not describe what they were seeing as "organisms" collected from various "environments," for the simple reason that neither term (or their closest French equivalents, *organisme* and *milieu*) came into common use among naturalists until decades later.[62] Nor did they see their observations and experiments as part of the scientific discipline of "biology" (*biologie*)—this too was a term that came into wide use only much later, having been independently coined by several naturalists, including the museum's own Jean-Baptiste Lamarck, only a few years before the *Géographe*'s arrival in Lorient.[63] Rather, naturalists were still working largely within the vocabulary and conceptual framework established by Buffon and other mid-eighteenth-century naturalists. As a result, they described what they were seeing as plants and animals gathered from diverse places around the world, which they understood as a subset of the many kinds of products of nature that fell within the ambit of natural history—a capacious field of activity that also included the study of fossils, minerals, climates, and peoples. Nonetheless, by the beginning of the nineteenth century, the museum's naturalists were already beginning to develop techniques of classification and experimentation on the premise that living beings, as distinct from other kinds of natural objects, could best be understood as organized bodies whose internal structure corresponded in some essential way to the external conditions they faced—a relationship that by the 1830s some naturalists would begin to describe in terms of "organisms" and their "environments."

To understand how and why new ways of studying life in environmental terms emerged during this period, it is necessary to understand the older ways that naturalists were responding to and building upon. In France, no eighteenth-century naturalist left a more important legacy than Buffon. Although many contradictions and inconsistencies can be found throughout his enormous body of writing, his explanations for the diversity of living forms and their ability to reproduce themselves over generations retained a powerful influence long after his death in 1788. More important, perhaps, than the specific answers he offered were the kinds of questions he posed—in particular, questions about how the complex organization of living bodies could be explained in terms of various forces and forms of matter interacting according to natural laws. Even though these questions were answered differently by the following generation of naturalists, they still owed much to Buffon's original framing.

Despite being a thoroughgoing materialist, Buffon did not think that the obvious complexity and dynamic nature of living beings could be explained as the product of mechanical forces acting on passive matter, as René Descartes had suggested a century earlier.[64] Instead, he argued that living beings were composed of a special kind of matter consisting of "organic molecules," which was governed by a distinct set of natural laws and which circulated from one living being to another without being changed in the process.[65] In order to explain in material terms—that is, terms that did not require the intervention of a supernatural or immaterial force—how living beings of a particular kind were able to "reproduce" themselves over time, he suggested the existence of "internal molds" that determined how organic molecules were incorporated into the bodies of living beings. It was these internal molds, unique to each species of plant or animal, that allowed living beings to propagate themselves through a process that was completely material but distinctly different from the kinds of transformations affecting nonliving matter.[66]

The concept of the "internal mold" also gave Buffon a means of explaining the influence of external conditions on variations among living beings of the same kind. Although he resisted the idea that one species could be transformed into another across generations, he did believe that species could "degenerate" when their conditions of life changed significantly.[67] The main mechanisms for such degeneration, in his view, were changes in climate and in the kinds of organic molecules available in a living being's surroundings. When the climate differed significantly from that to which the species was best suited, when the available organic molecules were less nutritive or in some other way inferior to those to which the species was accustomed, or when the species came under the influence of humankind, the result was a change in the species' original form.[68] This "degeneration," as Buffon called it, was not extensive enough to give rise to entirely new species, but it could explain why similar kinds of living beings were distributed across the globe in ways that suggested dispersion from a single center of creation. Over generations, the "general prototype" of each species was passed across generations via the internal mold, even as the actual embodiment of any particular plant or animal shifted in response to the conditions around it.[69]

Buffon's view of the role of interactions between the internal mold and the climates and other conditions it encountered sound similar to later conceptions of the relationship between organisms and their environments, but there are fundamental differences. For one thing, Buffon had no sense of a process of adaptation between a species and its surroundings. As the name suggests, "degeneration" was usually a negative process, one that led a species to reflect the characteristics of its surroundings rather than adapting to them.[70] Just as depriving an individual plant of sufficient water led to a smaller and weaker plant instead of a plant that was well adapted to arid conditions, so exposing a species to a different climate or set of organic molecules resulted in a variation of that species that was inferior to its original form. Closely linked to this idea was Buffon's conviction, shared by many at the time, that nature was arranged in a hierarchy of increasing perfection, with humanity at the pinnacle.[71] Any changes in a living being's form could, he thought, only move it downward along that scale. The absence of an idea of adaptation in Buffon's thought can also be seen in his well-known statement that "all that can be, is" ("*tout ce qui peut être, est*"). As a firm believer in divine omnipotence and the plenitude of nature, Buffon was convinced that the Creator had fashioned "a world of related and unrelated beings, an infinity of harmonic and contrary combinations, and a perpetuity of destructions and renewals."[72] In such a world of variation and change, it would be foolish to try to explain all organic forms in terms of adaptation.

Buffon's vision of a world replete with diverse forms of life that had been divinely created but were also shaped by their surroundings left him with an ambivalent and inconsistent position on the significance of each species' particular patterns of anatomical organization. On one hand, his interest in the ability of living beings to "reproduce" themselves across generations in recognizable form made him intensely interested in their anatomical structures and functions. On the other hand, if it was indeed true that all possible forms existed in nature, then there must necessarily be some forms—both entire species and parts of an individual organism's body—that were inferior to others or even useless. A bird's feathers clearly helped it fulfill the function of flying through the atmosphere, as did much of the rest of its body, including its hollow bones and its powerful wing muscles. Why, then, were there species of birds with hollow bones and feathers that were nonetheless incapable of flight? Such cases seemed proof that the organization of living bodies did not always have a meaningful relationship with their surroundings or way of life. Unable to believe that species changed gradually over time in ways that left traces of the adaptations of their evolutionary ancestors, Buffon had little recourse but to speculate about nature's endless variety.

Despite the ambiguities and contradictions of Buffon's writings, he suggested to naturalists of the following generation the importance of understanding how and why the organization of a living body was related to the conditions in which it lived, whether those conditions were defined in terms of climate, diet, or some other factor. Indeed, naturalists such as Lamarck and Cuvier—the former a student of Buffon's, the latter strongly influenced by him—made the question of functional organization one of the central preoccupations of their work.[73] Although the two men disagreed on the answers, they started from the same premise: that even the most extravagant or apparently useless organs resulted from a set of underlying organizational principles that could be discovered through the kinds of research they were pursuing at the museum. While each organism may well have been divinely created in perfect form, they thought, the purpose of its design was to enable it to survive under particular conditions, not to demonstrate divine omnipotence or nature's endless creativity—much less to convey a particular moral lesson to humanity, as the compilers of medieval bestiaries had sometimes suggested.

The generation of naturalists that led the museum after its founding in 1793 was also influenced by Buffon's conviction that life could ultimately be explained in material terms, although they largely abandoned his ideas of "organic molecules" and "internal molds." Rather than assuming that a living thing was distinguished from other kinds of objects by the kind of matter it was made of, they argued that it was distinguished by the way in which ordinary matter was organized, although they often disagreed about precisely how that process of organization worked. Lamarck, for instance, believed that ordinary matter assembled itself into self-reproducing, self-organizing bodies that constantly strove toward greater perfection in relation to the circumstances they faced. It was this process of organization, he argued, rather than any specific property of the matter they were made of, that distinguished living things from nonliving things.[74] Cuvier, by contrast, saw no reason to believe that species were inherently self-improving, or indeed that they could change significantly over time. Nonetheless, he agreed that living beings were best understood as organized bodies in which each part contributed to the survival of the whole, and that it was this form of organization, rather than the fact that they were constituted by a special sort of living matter, that made them alive.[75]

The idea that the unique properties of living beings were determined by the structural and functional organization of their bodies had implications for practical research as well as for general theories of nature's order. For Cuvier, for example, it was clear that all parts of an organism were so intimately related to each other and to the organism's way of life that changes in any one organ would necessarily be reflected throughout the entire body.[76] If birds were to be able to fly under the conditions of Earth's gravity and atmosphere, for example, their skeletons would have to be both light and strong enough to be lifted by their powerful wings without breaking under the force they exerted. Formalized as the "correlation of parts," this principle helped Cuvier reconstruct extinct species on the basis of fragmentary fossil evidence. It also suggested that nature's creativity was bounded, since any living being that was not well organized in relation to what Cuvier described as its "conditions of existence"—the necessary relations between all its constituent parts in light of its role in nature—would, by definition, be unable to survive.[77] "As nothing can exist if it does not assemble the conditions that render its existence possible," he wrote, "the different parts of each being must be coordinated in such a manner as to render possible the being in its totality, not solely within itself, but in its relations with those that surround it."[78] For that reason, it could hardly be true that nature was replete with both organized and disorganized living forms, as Buffon had argued half a century earlier. Rather, only those forms would exist that allowed living beings to survive under the conditions found on Earth.

While Cuvier rejected the idea that a species could transform into another species over time, his colleague Lamarck saw the drive toward perfection—both within the lifespan of an individual plant or animal and across generations—as fundamental to the organization of living bodies. In some ways, Lamarck's theory of species transformation bridged Buffon's views and those of later "evolutionists" of the nineteenth century such as Charles Darwin.[79] In the case of plants, he argued for something like Buffon's explanation—namely, that any observable changes in a species were due primarily to the direct effects of climate, mechanical pressure, or the incorporation of particular kinds of food or other matter into their bodies.[80] These direct effects did not necessarily make the plant species more adapted or more "perfect"; they were simply consequences of the changed conditions in which it lived. For animals, however, the case was radically different. In order to accomplish basic functions of life such as respiration, digestion, locomotion, and reproduction, Lamarck argued, animals had to be capable of changing their structure dynamically in relation to surrounding conditions. Their responses to surrounding conditions were not uniform, as those of plants were; rather, the effect of climate, diet, and other factors depended on the organization and aims of each particular kind of animal. Over time, as animals sought to perfect themselves in relation to their changing surroundings, he believed, they would pass those changes on to their offspring, leading to the observable diversity of the natural world.[81]

More broadly, the focus on processes of organization changed the way naturalists thought about the influence of external conditions on living beings. To begin with, it diminished the appeal of theories that attributed a direct mechanical influence to surrounding conditions—whether via "organic molecules" or some other means—since naturalists now believed that any such influence would be mediated by the distinctive way in which each organism was organized. The belief that external influences could be viewed as material forms that left impressions on living beings, including Buffon's idea of an "internal mold," was gradually replaced by the idea of a functionally integrated organism seeking to survive and reproduce under certain conditions whose significance varied from organism to organism. This functional perspective also had implications for naturalists' understandings of "climate" and "place." Since the effect of climate was now understood to depend on the organization of each living body, it was no longer possible to make general claims about the effects of raising or lowering the temperature. Similarly, if each organism in a particular place was affected differently by prevailing conditions, then there was nothing that could be said in general about the character of plants, animals, and humans in specific geographical locations. Climate and place thus declined in importance as ways of explaining the character of living things, becoming instead subordinate aspects of something broader that did not yet have a name.

The generation of naturalists after Cuvier and Lamarck took this line of thought even further. In his 1822 *Treatise on Animals* and in a series of lectures on physiology delivered between the late 1820s and early '30s, for example, Cuvier's student Blainville sought to explain life in terms of organisms, the external forces and materials that impinged on them, and what he called the "envelope" of organs that served as interfaces between the two.[82] An organism's effective world, he argued, was determined by the organs of sense and action arrayed upon its surface, whose precise character in turn reflected the organism's inner structure and needs as well as the external conditions under which it developed. From this perspective, he argued, the "natural history properly speaking" of any given species consisted of "the different manners in which these combinations of organs . . . act on external circumstances to nourish and propagate themselves."[83] This view of life also suggested the need for new methods of research. In order to make meaningful claims about the relationship between a living being and its surrounding conditions, Blainville argued, naturalists would need to understand the precise physical and chemical processes that mediated between an organized body, on the one hand, and the external circumstances that it encountered, on the other.[84]

From the 1790s to the 1830s, most discussions of such issues were conducted without using the words *organism* or *environment*. Instead, French naturalists deployed a wide range of terms to describe the conditions in relation to which living beings were organized, each with its own connotations and shades of meaning. These included *milieu*, usually understood in the limited sense of physical medium in which an organism lived (such as air or water), but also phrases such as *external circumstances, surrounding influences,* and many others. By the 1830s, the basic idea that lay behind this proliferation of terms was taken for granted by most French naturalists, many of whom had been trained by Cuvier, Lamarck, Geoffroy, Thouin, and others from the Muséum d'Histoire Naturelle. At the end of the decade, the philosopher Auguste Comte, a student and friend of Blainville's and a close follower of the work of Cuvier and other Parisian naturalists, redefined *milieu* to mean "the total set of external circumstances, of whatever kind, necessary to the existence of each specific organism," including the human organism.[85] In doing so, he offered a singular term for a concept that had been developing under a variety of labels since the end of the eighteenth century. This was a concept that had become compelling not simply because it described nature accurately, but also because the museum's naturalists had built an institution and a set of practical techniques that made it visible for all to see.

BEYOND THE MUSÉUM D'HISTOIRE NATURELLE

If the concepts of "organism" and "environment" had remained tightly linked to the naturalists at the museum and their research methods and theoretical preoccupations, it seems unlikely that these concepts would have been as widely adopted as they were by the middle of the nineteenth century. Instead, as naturalists developed new techniques for studying the relationship between organisms and their environments, they gradually came to recognize the limits of the Muséum d'Histoire Naturelle and institutions like it. They began seeking out new venues where they could explore alternative methods of research, collaborate with experts in other domains, and explore the practical implications of their theories on a grander scale than was possible within the confines of the museum and its garden and menagerie. The result was the beginning of a process that would continue over the following two centuries in an even wider range of situations and with even more fundamental adjustments in the ways that environments were conceived—namely, a process of adopting and reshaping the concept of environment to suit new needs and circumstances.

One path away from the museum led toward the physiology laboratory, which would become the site of many of the most striking scientific advances in the emerging field of biology in the nineteenth century. In at least one important case, this turn toward the laboratory was motivated partly by institutional politics. After a falling-out with Cuvier in 1817, Blainville was effectively banned from the use of the collections for a number of years. Working under this not insignificant constraint, he concentrated his efforts on studies that did not require access to large collections but instead could be carried out with only a few anatomical specimens.[86] He began to explore the use of methods and venues beyond those of museum-based comparative anatomy, including those of medicine and physiology, at the same time that he continued to build on the idea of functional organization advocated by his mentor and rival Cuvier. In his own work, he persisted in carefully dissecting and describing the internal structure of the organism in relation to its conditions of existence, but he and his students and followers also began to conduct experimental tests of the physiological functions of the organs that they believed mediated between an organism's internal structure and its external conditions.[87]

In this way, Blainville's more or less accidental exclusion from the museum in the late 1810s and '20s encouraged him to take his interest in the functional organization of living beings in a new direction. It shifted his attention from the kind of comparative anatomy advocated by Cuvier to new forms of physiological experimentation that aimed not merely to show that the form of an organism corresponded to its role in nature but also to reveal the precise physical and chemical mechanisms by which that correspondence was mediated. One of those who followed him in pursuing this approach was Jean-Louis-Maurice Laurent, a protégé of Thouin who was deeply familiar with the methods and ideas developed at the museum at the beginning of the century but also eager to move beyond them through the methods of experimental comparative physiology advocated by Blainville.[88] As editor of the *Annales Françaises et Étrangères d'Anatomie et de Physiologie*, a journal founded in 1837 to champion the new approach, Laurent published both his own and others' experiments on the influence of surrounding conditions on the development and functioning of living beings. In the first issue of the journal, for example, he reported that he had been able to suspend and resume the development of a certain species of mollusk by submersing its embryos alternately in aerated and non-aerated water, a result that indicated the critical relationship between external conditions and internal developmental processes.[89]

Another path away from the museum led toward studies of the relationship between organisms and their environments in the places where those organisms were originally found. The work of Alexander von Humboldt, probably the museum's most renowned correspondent, was particularly influential in this regard. Rather than assuming that climatic zones were sufficient to explain the distribution of plants across the surface of Earth, Humboldt sought to characterize the conditions under which plants grew in terms of various physical factors—temperature, precipitation, sunlight, air pressure, soil chemistry, and so forth—that could be precisely quantified, tabulated, and mapped. He thereby contributed to the emergence of the concept of environment as distinct from both place and climate that had been articulated in much of the work conducted at the museum from the late eighteenth century onward.[90] He also adapted the concept to his own purposes, focusing much more on the quantification and mapping of variations in physical parameters than the naturalists of the museum had done.

The influence of these methods and concerns can also be seen in the work of the Swiss botanist Augustin Pyramus de Candolle, who had spent a decade in Paris studying and working under Lamarck after the French annexation of Geneva in 1798 and who was a close follower of Humboldt's work.[91] In addition to working with the collections of the Muséum d'Histoire Naturelle and its garden, Candolle had extensive experience collecting living plants in the field, including carefully observing the conditions under which they grew. In an 1820 essay on botanical geography, he proposed a conceptual distinction between the "habitations" of plant species, which he defined as their distribution across Earth's surface, and their "stations," which he defined as the climate, soil, and other conditions under which particular varieties of plants were capable of growing. "Certain plants, according to their organization, have need of certain conditions of existence," he wrote; "one cannot live where it does not find a certain amount of salt water; the other where, at a particular time of year, it does not have such an amount of water or such an intensity of sunlight, etc."[92] The best place to determine those conditions, Candolle argued, was in the field, by traveling researchers who gathered not only specimens but also "precise and varied details on the stations and habitations of plants."[93] In this way, he helped bring the emerging concept of environment beyond the walls of the museum and into the field.

The idea that life could best be understood in terms of the relationship between organisms and their environments also inspired French naturalists to propose ambitious breeding projects that were not easily accommodated within confines of the garden or the menagerie. In the menagerie in particular, efforts to acclimatize exotic species to Parisian conditions had been constrained by limited resources and conflicting priorities. Especially problematic was the museum's effort to maintain a diverse collection of novel animals that would attract the interest of the public and enrich the galleries of comparative anatomy after their deaths, which discouraged the keeping of numerous specimens of any one kind of animal—precisely what was needed to support an effective acclimatization program. In any case, keeping any number of animals alive and reproducing in the cramped quarters of the menagerie proved to be a virtually insurmountable challenge. As a result, despite Frédéric Cuvier's experiments with hybridization in the early decades of the nineteenth century and his outspoken advocacy of acclimatization projects from the 1820s onward, few long-term experiments in acclimatization were conducted during his term as keeper.[94]

By the 1830s, it had become clear that success in acclimatizing exotic animals to France or to its colonies, including the newly conquered territory of Algeria, would require larger facilities dedicated specifically to the task. After Frédéric Cuvier's death in 1838, management of the menagerie was put in the hands of Geoffroy's son Isidore Geoffroy Saint-Hilaire, a firm believer in the possibility of species transformation. In a 1861 report on the *Acclimatization and Domestication of Useful Animals*, he argued that the "character of organized beings, fixed for each species as long as it perpetuates itself within the same circumstances, is modified if the ambient circumstances come to change."[95] Appearing two years after Charles Darwin's *Origin of Species* introduced a new model of evolution by natural selection, the report reflected an increasing acceptance of the idea that species could change radically in response to the external conditions they faced, which included other living beings.[96] It also had direct implications for acclimatization. Over the course of several generations, Isidore Geoffroy argued, exotic animals accustomed to warmer, drier climes could be protected long enough to adapt to Parisian conditions; eventually, they would develop thicker pelts and hardier constitutions that would be passed on to their offspring as heritable traits. As part of his advocacy for such experiments, Geoffroy established the Société Zoologique d'Acclimatation in 1854, as well as a farm dedicated solely to acclimatization, the Jardin Zoologique d'Acclimatation, in 1861.[97] Whereas the menagerie had helped make the distinction between the organism and its environment visible, the Jardin d'Acclimatation sought to apply it on a large scale.

In the middle decades of the nineteenth century, naturalists associated with the Muséum d'Histoire Naturelle also began to consider how the methods that they had developed might be applied both to individual human beings and to entire human societies. In their wide-ranging 1847 volume on the *History of the Sciences of Organization and Their Progress, as the Foundation of Philosophy*, Blainville and François-Louis-Michel Maupied argued that the role of natural history within the grand scheme of human knowledge was to shed light on the chain of being that began with the lowest organisms and rose gradually toward its pinnacle, humanity.[98] As one ascended this chain or series, they argued, one observed an increasingly nuanced, complex, and extended set of processes mediating between the organism and its surroundings. In understanding this relationship, zoology had a special role to play, according to Blainville and Maupied. Seen not as the science of animals but rather as the science of relations between organized bodies and their surroundings, zoology became the highest and most refined means through which humanity could improve its own relationship to its environment. Its scope was thereby expanded to include "all the sciences that relate to man and to animals, including that of their education and their government."[99]

These kinds of ideas helped create a bridge between the study of organisms and their environments and the reform of human societies, one that would give the concept of environment a tremendously broad scope in subsequent decades, even as it continued to be adapted to specific circumstances and purposes. In the French context, the central role given to *milieu* in Comte's system of "positive philosophy" was perhaps the most explicit expression of this broadening of the concept beyond its roots in natural history. Like Blainville, Comte understood life as being arrayed in a series from least to most complex, with humanity at the top. As one ascended the series, organisms became both more sensitive to the conditions surrounding them and more capable of changing those conditions.[100] In light of humanity's unique combination of vulnerability and power, Comte sought to develop an ambitious system—first a systematization of all the sciences, and later an all-encompassing secular religion of positivism—that would improve humanity's relationship with its surroundings. Among the many people influenced by this explicitly "environmental" view of the cosmos was the British philosopher Herbert Spencer, who —drawing on a loose translation of Comte's *Cours de Philosophie Positive* by Harriet Martineau—was largely responsible for introducing the term *environment* into English-language scientific discourse in the second half of the nineteenth century.[101]

*

In an 1808 report to Napoleon on the state of the sciences in France, Cuvier compared life to a "continual vortex, of which the direction, as complicated as it is, remains constant, as does the type of molecules that are carried along with it, but not the individual molecules themselves."[102] Life was not a specific kind of matter, in other words, but a way of organizing matter. In one sense, the Muséum d'Histoire Naturelle, too, was an organized vortex, pulling in specimens and observations from around the world and structuring them to reveal the principles that governed the relationships between organisms and their environments. That vortex did not operate in a vacuum, however. Instead it was highly dependent—as Cuvier's report to Napoleon also suggested—on the efforts of the French state to expand its imperial reach both within and beyond Europe. Without it, Cuvier and his colleagues would likely have had many fewer specimens to study, far fewer resources to do so, and less motivation to understand the conditions necessary for organisms to thrive far away from their places and climates of origin. In this sense, the emergence of the idea that life can best be understood in terms of the interactions between organized bodies and the conditions necessary for them to flourish cannot be understood without taking the history of European empires into account.

The image of the Muséum d'Histoire Naturelle as an organized whirlpool also suggests the importance of physical setting and practical techniques for the emergence and adoption of new concepts such as "organism" and "environment." In the course of their efforts to understand and manipulate the conditions of existence of diverse animals and plants from around the world, museum naturalists became intimately connected to certain aspects of the world and not to others. We cannot understand their changing relationship to the natural world solely by paying attention to the theories and concepts they expressed in books and in journals such as the *Annales*, as if writing alone were capable of bringing new ways of being into the world. Rather, we need to look at the specialized practices, associated with particular artifacts and settings, that allowed these naturalists to observe and manipulate the relationships between organisms and their conditions of existence. These practices of collecting, classifying, and experimenting with living and preserved specimens motivated naturalists to develop new concepts, and helped them transform the everyday worlds around them in ways that made concepts such as "organism" and "environment" increasingly plausible.

Examining a community of people who adopted these concepts after long making do without them—indeed, one of the first, if not the very first, communities to do so—also helps us understand how the idea of "environment" (or *milieu*, in the French case) differed from what might seem in retrospect like closely allied concepts, such as "climate" and "place." There is no universal answer to this question, since each community is different, but there are patterns that can be identified in each specific case. For French naturalists at the beginning of the nineteenth century—some of whom had been trained by Buffon and all of whom were influenced by his work—the idea of the functional organization of living beings in relation to what Cuvier called their "conditions of existence" and "role in nature" was essential. More an overarching question than a predetermined answer, this view of the living world focused their attention on the processes that transformed unorganized matter into organized living bodies. Through such processes of organization, they came to believe, each living being encountered the world in its own way.

Finally, looking at the development of techniques, practices, and concepts of "organisms" and "environments" at the museum from the late eighteenth century to the mid-nineteenth century helps us see how a concept that originally emerged within one specialized setting could be translated and adapted to serve in other settings—sometimes of a radically different character. After the mid-nineteenth century, the museum's status as Europe's preeminent institution of natural history faded, as indeed did the reputation of natural history itself in comparison to new scientific disciplines such as biology and geology. By that point, however, the concept of environment that the museum had incubated had made its way, with various refinements and adaptations, into a wide range of other fields, from the kind of botanical geography advocated by Humboldt and Candolle to the practical experiments in acclimatization pursued by Isidore Geoffroy to the grand projects of systematization and social reform advocated by Comte and Spencer. The following chapters explore a variety of communities of scientists, reformers, and activists that carried the concept even further afield, in each case with the help of specific techniques, practices, and settings. The next chapter turns to British physicians of the mid-nineteenth century who adopted new methods of medical statistics in the hope of preventing or treating disease in their nation's expanding empire.

CHAPTER 2

Environments of Empire: Disease, Race, and Statistics in the British Caribbean

In its milder forms, yellow fever's symptoms include fever, headache, nausea, and fatigue, none of which clearly distinguish it from a number of other diseases. At its severest, however, the telltale signs are clear: high fever, jaundice, bleeding, shock, and organ failure, almost always ending in death. In late 1866, these dreaded symptoms of the disease sometimes known as "yellow jack" began to appear among sailors, soldiers, and civilians in Jamaica.[1] First noticed among sailors and merchant marines at Port Royal who had most likely been exposed to the disease by one of the many ships calling at Kingston from ports elsewhere in the Caribbean, it spread rapidly across the island.[2] By February 1867, it had reached troops stationed at Up-Park Camp, the main barracks for British soldiers at Kingston. The pace of the epidemic seemed to slow in March and April, but it quickened again in the following months. Its impact was felt especially in coastal towns but also reached the highlands, including Newcastle, a British hill station located above 3,800 feet in the Blue Mountains. By the time the epidemic had run its course at the end of 1867, official army statistics reported the deaths of thirty-one noncommissioned officers and enlisted men, five officers, and five women among the military population, along with many more of the island's civilians.[3]

At a time when the causes of yellow fever were disputed and treatments were largely ineffective, British authorities responded to the disease primarily by relocating soldiers to parts of the island that it seemed unlikely to reach—in particular, high-altitude areas like the Blue Mountains. In February, three soldiers diagnosed with fever at Up-Park Camp were sent to Newcastle in the hope that the altitude and climate of the station would do them good. Although all three died within days of their arrival, army commanders continued to believe in the protective effect of higher altitudes. After the death of another soldier at Up-Park Camp in April, they ordered most white troops to be moved to Newcastle, although troops of African descent, whom they assumed to be largely immune to the disease, were left at coastal stations. Not only did soldiers and sailors who had already contracted the disease in the lowlands continue to sicken and die, however, but even those in seemingly perfect health were struck down, including members of the 84th Regiment who had made the short trip to Newcastle immediately after arriving in Jamaica from Malta.[4]

Disturbing as it must have been to those who lived through it, the 1866/67 epidemic of yellow fever was neither the first in Jamaica nor the deadliest. The disease had reached epidemic proportions on the island as recently as the mid-1850s, and earlier in the century it had regularly carried off a significant proportion of British troops even in years without epidemics.[5] Other diseases such as cholera and malaria also took a heavy toll. Nonetheless, the 1866/67 yellow fever epidemic was troubling to army officials because of its extensive impact at Newcastle, which the army had relied on as a refuge from lowland fevers.[6] Its faith in the protective effect of high altitudes shaken, the Army Medical Department appointed a commission to document the course of the epidemic and investigate the factors that affected it.[7] Submitted in September 1868, the final report of the commission included a map of Jamaica, a description of the island's topography and climate, a survey of the health of the population immediately before the outbreak, and accounts of every known or suspected case.[8] It focused especially closely on Newcastle, providing detailed information about its climate, architecture, layout, sanitation, and surrounding vegetation. Complementing these maps and descriptions were statistical tables displaying in quantitative terms the variable impact of the epidemic at different places and on people of different kinds.

The commission's use of maps and statistics to study the epidemic reflected changes during the preceding decades in how British physicians had come to understand the relationship between external conditions and the health of individuals. Taking advantage of an expanding British infrastructure of military stations during a period of relative peace between European empires, army physicians had gathered masses of data on disease prevalence among soldiers and sailors stationed around the world. They used these data to test in quantitative terms some of the generalizations that had emerged from the long British tradition of neo-Hippocratic medicine—that is, medical theories and practices inspired by the writings of the ancient Greek physician Hippocrates on the effect of "airs, waters, and places" on health.[9] While the findings of these

SKETCHES AT NEWCASTLE, JAMAICA

FIGURE 4. Sketches of the Newcastle hill station published in the 1870s, depicting among other things a coffin containing a "victim to Yellow Jack." (Reprinted from "Sketches at Newcastle, Jamaica," *Graphic*, April 20, 1878, 400.)

medical statisticians confirmed some of the claims of neo-Hippocratic physicians, they disproved others or showed that they were valid only under certain circumstances. Those disputed claims included the idea that high altitudes offered sure protection against diseases such as yellow fever. High altitudes were indeed associated with lower prevalence of such diseases but did not prevent them entirely. The interactions between bodies and their surroundings were more complex than could be captured by altitude or temperature alone, albeit in ways that medical statisticians acknowledged were still beyond their understanding.

The concept of environment was adopted and adapted by British physicians in the last decades of the nineteenth century to make sense of these complexities. It expressed an emerging consensus that disease could best be understood as the result of a disordered relation between a functionally integrated human body and the external conditions that it required to survive and flourish, rather than as the result of "humors" in the human body that could be thrown out of balance by influences from its surroundings. This shift in perspective was linked to an intensified focus on personal hygiene, waste disposal, and other sanitary measures, as well as to research on the role of nonhuman organisms—particularly bacteria, viruses, and insects—in causing and spreading infectious diseases. Yellow fever outbreaks, for example, were no longer seen as the product of mysterious interactions among place, climate, season, and constitution, but rather as the biological actions of a microscopic disease agent transmitted to human bodies by mosquitoes that flourished under particular conditions. In this view, in contrast to the older neo-Hippocratic medicine, there was nothing inherently unhealthy about any particular climate or place; rather, disease emerged when living beings organized in particular ways encountered physical and biological environments for which they were poorly suited.

In the context of the nineteenth-century British Empire, physicians' emerging environmental understanding of health was heavily inflected by preoccupations with racial difference. In Jamaica, for example, the abolition of slavery in 1838—just over three decades after the abolition of the transatlantic slave trade—had led to decades of struggle between white plantation owners who sought to maintain their rule over the island and black Jamaicans who demanded economic opportunity and political enfranchisement. In this context, claims about the differential biological vulnerability of various races to tropical diseases had profound political implications. Both imperialists and anti-imperialists marshaled medical statistics and biological theories of racial difference to argue that the bodies of Africans were organized in such a way as to thrive under the conditions of plantation labor in the Caribbean, while the fragile bodies of Europeans required special protections to survive in tropical environments. Influenced by the social and material conditions they encountered at the front lines of the imperial project, British medical officers adapted the concept of environment to their own unique circumstances and aims in ways that preserved existing racial and social hierarchies. Even after the British presence in the Caribbean faded and new theories of "germs" were developed to explain disease, these hierarchies continued to shape the way health was understood.

COLONIZING

Just as naturalists at institutions such as the Muséum d'Histoire Naturelle in Paris benefited from the global expansion of European empires from the fifteenth century onward, so did physicians benefit from opportunities to observe how people of different types sickened or thrived as they moved from one region of the world to another. For British physicians in particular, the growth of their nation's empire in the nineteenth century and its reorientation toward tropical regions of Asia, Africa, and the Americas in the wake of the loss of most of its North American territory offered expanding horizons and new sources of data. People, diseases, and information circulated at accelerating rates as new colonies, many of them in tropical climates, were acquired and linked into a global network of trading routes, naval ports, and military stations.[10] As integral parts of this network, British medical officers were well positioned to observe the sometimes deadly results of displacing human beings from their native climates, places, and communities and subjecting them to a variety of new conditions. Their observations, records, and experiments— some intentional, some accidental—gradually provided the foundation for a new understanding of sickness as the product of a disordered relationship between the human organism and its environment.

The world they observed had already been reshaped over centuries by the movements of people and diseases that had been catalyzed by European exploration and colonization. In populations that had not previously been exposed to highly virulent Old World diseases such as smallpox and yellow fever, the absence of immunity acquired through exposure during childhood had contributed to extraordinarily high rates of mortality among indigenous populations.[11] The effects were often exacerbated by the social and ecological disruptions that resulted from European colonization, which compromised the ability of indigenous communities to fight off infections and care for the sick.[12] While some of these disruptions were intentional—indeed, some were explicitly genocidal—others were more or less accidental. In the Caribbean, for example, the conquest, enslavement, and displacement of the indigenous Taíno population was followed by changes to the landscape that unleashed new sources of disease, including deforestation and the proliferation of pits, tanks, and other opportunities for stagnant water to gather on and around plantations. Although Europeans at the time were unaware of it, these changes expanded the habitat for the mosquito species that transmitted yellow fever, a disease most likely introduced to the Caribbean by slave ships in the sixteenth or seventeenth century.[13] These newly arrived diseases had major demographic consequences that compounded the direct effects of European colonization. By the time the British took Jamaica from the Spanish in the mid-seventeenth century, for example, the Taíno population had been nearly eliminated by military conquest, social disruption, and harsh labor conditions, as well as by smallpox, yellow fever, and malaria.[14]

Differential vulnerability to disease also shaped the demographics of the Caribbean through the transatlantic slave trade, which brought millions of West Africans to labor on sugar, cotton, indigo, and coffee plantations on Jamaica and other fertile Caribbean islands and coastal areas from the sixteenth century to the beginning of the nineteenth century.[15] Coming from places where smallpox and yellow fever were endemic, West Africans forced to labor on Caribbean plantations proved less vulnerable to those diseases than the region's indigenous peoples. They benefited from immunity acquired during childhood exposure to disease as well as from healing traditions—both their own and those they learned from indigenous Americans—that did not cure the disease but sometimes eased its symptoms and, in any case, did less harm than most European forms of treatment.[16] Over time, the proportion of people of African descent in the Caribbean continued to grow even as indigenous populations collapsed. Meanwhile, the population of European settlers grew only slowly, in part because plantation owners avoided living in the region for fear of infection.[17] The result was that by the time slavery was abolished in Jamaica, people of African descent made up approximately 84 percent of the population of the island, with the rest consisting mainly of a small ruling class of white plantation owners, merchants, and colonial officials.[18]

Tropical diseases and differential vulnerability to them—due not to inherent biological differences among races but rather to variations in exposure and in the social capacity to cope with disease—also affected the ability of European armies to conquer territories, establish permanent colonies, and suppress rebellions. While most European communities were familiar with the horrors of smallpox, those in Britain and other northern regions of the continent were spared diseases such as yellow fever and malaria that were carried by species of mosquitos that could not survive the northern winters. Combined with the ordinary sanitary problems of all military encampments, the exposure of immunologically naive populations of European soldiers to tropical conditions resulted in high rates of morbidity and mortality. At the turn of the nineteenth century, for example, British and French attempts to suppress the Haitian revolution failed largely because of disease.[19] Among the British troops sent to what was then the French Caribbean colony of Saint-Domingue between 1793 and 1798, approximately two-thirds of died of yellow fever and other diseases.[20] Similarly, the French attempt to retake the colony in 1802 was stymied by the loss of 35,000 to 45,000 soldiers out of a total force of 60,000 to 65,000, most of them to yellow fever.[21] At the same time, the disease left the rebels, many of whom had probably acquired immunity through childhood exposure, largely unscathed—an advantage that Toussaint Louverture and other Haitian leaders strategically exploited over the course of their successful revolt.[22]

Even when tropical diseases did not prevent Europe's imperial states from acquiring colonies or suppressing rebellions as they did in Haiti, they placed a heavy burden on their armies and navies. Disease consequently remained a central preoccupation of military commanders, medical officers, and colonial administrators throughout the nineteenth century in times of both war and peace. In Jamaica, for example, the British lost about 8,000 men to yellow fever during the Napoleonic Wars even though the island never became a site of active combat, and annual losses to diseases of all kinds among soldiers stationed there remained over 10 percent through the 1830s.[23] Yellow fever remained a scourge of British troops until the last third of the nineteenth century. In 1840/41, for example, an epidemic carried off 282 soldiers in a force of 1,153—a mortality rate of nearly 25 percent.[24] In subsequent decades, deaths from disease among British troops in Jamaica declined dramatically, probably due to sanitary reforms and the use of high-altitude stations such as New-castle.[25] Nonetheless, disease mortality continued to exceed mortality from any other cause, and death rates at military posts in tropical regions remained higher than rates in the British Isles or temperate North America through the end of the century.[26] Despite the absence of yellow fever in India, for example, British troops there suffered enormous casualties in the first half of the nineteenth century from diseases such as malaria, typhoid, dysentery, and cholera.[27] In 1863, an official report calculated that more than 5,800 of the approximately 70,000 British soldiers stationed in India—that is, more than

8 percent—were in hospital beds at any given time, while more than 4,800 died of disease annually.[28]

While the accuracy of such numbers should not be overestimated, the fact that statistics are available at all is a product of changes in the British Empire in the nineteenth century that encouraged British physicians to study in quantitative terms the relationship between an individual's health and the external conditions he or she encountered.[29] One major factor was the shift in the British Empire's center of gravity from temperate North America to the plantations of the Caribbean and the new colonies in India, East Asia, the Middle East, Africa, and Oceania.[30] As the geography of empire changed, physicians' views of tropical disease vulnerability changed along with it. Rather than celebrating the comparative healthiness of colonists of North America in relation to indigenous populations that had been devastated by smallpox and other Old World diseases, they became concerned about the fragility of British soldiers and sailors in tropical regions where the indigenous inhabitants seemed to thrive despite the presence of virulent diseases such as yellow fever and malaria.[31] At the same time, British physicians benefited from the increasingly formal and bureaucratic institutions of the empire. In the decades following the Napoleonic Wars, the newly established Army Medical Department sought to identify the causes of disease among British troops through a project of statistical record-keeping.[32] This project took advantage of Britain's expanding imperial infrastructure of naval ports, army encampments, hill stations, and medical staff, which facilitated the gathering of detailed accounts of morbidity and mortality rates for populations of men whose characteristics could be precisely determined and whose living conditions were tightly regulated.

The implications of such studies went well beyond concerns about the health of soldiers and sailors. In Jamaica, for example, they shaped the way British colonial administrators and military commanders responded to the political and economic struggles that followed the abolition of slavery. Along with shifts in the global market for cane sugar (the island's main export), emancipation undermined the economic and political power of the white plantation owners. In its wake, they sought new means of defending their position of dominance, even as the island's black peasants and political leaders pressed the colonial government to expand their rights as British subjects.[33] Exacerbated by drought, these tensions reached their violent climax in the Morant Bay Rebellion of 1865, which began with a peaceful demonstration by black peasants, developed into an armed rebellion, and ended with the massacre of hundreds of Jamaicans by British troops.[34] The aftershocks of the rebellion and massacre were still being felt when the medical commission launched its investigation of the yellow fever epidemic of 1866/67 and shaped the way its findings were interpreted.[35] If the health of British troops could not be protected even with the help of hill stations such as Newcastle and the most advanced medicine of the day, the troops' ability to suppress rebellions such as the one that had taken place just two years earlier would be severely compromised. More broadly, if Europeans proved to be inherently and irremediably vulnerable to tropical diseases, the future of the empire throughout the tropics would be thrown into question.

By attending to the social and material conditions of health in Britain's tropical empire, we can see how a single process—in this case, the colonization and exploitation of Jamaica and other Caribbean islands—affected both the world that physicians were trying to understand and the methods available for them to study it. Physicians back in the British Isles were also trying to understand the causes of disease, but they did so with a very different set of resources and motivations. As the country industrialized, many of them turned their attention to the working-class masses concentrating in major centers of industry and trade, whose notoriously poor health was simultaneously a social and an economic problem. Their findings and theories resembled in many respects those developed by British medical officers in the tropics around the same time—and indeed, they were tightly connected to them—but they also differed in critical ways. Unlike in the factories of Manchester, Liverpool, or London, the plantations and ports of Jamaica and other Caribbean islands brought people from Africa, the Americas, and Europe together under conditions that facilitated the spread of deadly diseases. Thus, in comparison to their colleagues in the British Isles, medical officers in the tropics were far more concerned with the relationship among racial types, external conditions, and disease.

COMPARING

In the period from the end of the Napoleonic Wars in the 1810s to the eruption of the Crimean War in the mid-1850s, British medical officers exercised little influence over the living conditions or the behavior of the soldiers and sailors they treated. Their prophylactic recommendations were viewed as costly, inconvenient, and ineffective by army commanders and navy captains, and were consequently mostly ignored. Usually called in to treat disease only as a last resort, the treatments they had to offer were rarely successful. This was particularly true for yellow fever, which in its severe form continues to have a high mortality rate today and which in the early nineteenth century was often

treated with purgatives, bloodletting, or preparations of cinchona bark, each
of which had passionate advocates and none of which was capable of reliably
changing the course of the disease.[36] Despite or even perhaps because of these
disadvantages, medical officers stationed in the tropics took advantage of their
opportunities to observe the relationships between disease outbreaks, the conditions under which soldiers and sailors lived, and the personal characteristics
of those affected. Although they did not yet describe health in terms of the relationship between "organisms" and their "environments," they gathered a rich
store of observations that would later serve as the foundation of an account of
health rooted in a biological understanding of the human organism and the
conditions it required to thrive.[37]

One of the few advantages that medical officers stationed at British military
posts in the tropics had over their colleagues back home was the opportunity
to observe and compare the health of large groups of men of various races,
origins, and histories of exposure to tropical conditions. Partly because of the
high mortality rate of British troops in the tropics, new infusions of troops
were constantly arriving, which meant that at most times a medical officer
could observe soldiers who had arrived within the previous months as well as
those who had been resident for many years or even decades. In Jamaica in the
early 1830s, for example, about one out of every seven soldiers present at any
given time had arrived within the previous two years.[38] Physicians serving in
the tropics also gained numerous opportunities to observe the health of people
of various races. In the Caribbean, for example, the British army included
troops of both European and African descent. The number of white troops
stationed in Jamaica in the two decades after Waterloo ranged from 2,000
to 4,000; at the same time, there were also a large number of black troops in
Jamaica who were stationed separately and seemed to suffer from a somewhat
different range of maladies.[39] When an epidemic broke out, physicians were
able to compare the relative vulnerability of each of these groups and draw
conclusions about the factors predisposing them to disease.[40]

British medical officers also had many opportunities to observe the health
effects of local variations in weather, topography, and other conditions, all of
them "tropical" but nonetheless distinct. While most military stations in the
Caribbean as elsewhere were located close to coastlines or rivers for reasons
that were both logistical and strategic, their surroundings varied widely from
station to station. All the islands in West Indies could be characterized as having a tropical climate, for example, but there was wide variation in topography
and weather from island to island. Some were dry, flat, and exposed to strong
winds, while others were wet, thickly vegetated, and sheltered. Even within a

single island such as Jamaica, the conditions in lowland coastal areas differed dramatically from those in the highlands, while the presence or absence of marshes, forests, or rivers close to military stations could fundamentally transform the soldier's experience of the tropics. As one observer noted of Jamaica's highly variable landscape, despite its "tropical" location, "almost any variety of climate may be procured."[41] During times of active combat, moreover, a single body of troops might move from one station or encampment to another that was situated in the midst of quite different conditions.

In addition to comparing the health effects of geographical variations across military stations, physicians also had many opportunities to observe the role of variations in clothing, housing, and sanitation in either reducing or exacerbating the effects of climate and place. Most soldiers at any given station were required to wear a standard outfit, but the nature of that outfit differed across time and place in relation to local circumstances, available materials, and the preferences of individual army commanders. Usually intentionally different from the clothing worn by local civilian populations, these outfits were often suspected of promoting or hindering adaptation to the challenges of tropical climates for European bodies. Flannel, for example, was deemed by the physician Robert Jackson—who spent several long periods in the West Indies between the 1770s and the 1810s—to be useful for soldiers in Europe but harmful in the tropics, where he thought it led to excessive heat and perspiration.[42] Housing also varied significantly across military stations and even within a given station, and physicians sought to determine whether providing soldiers with insufficient sunlight, ventilation, or space contributed to disease outbreaks. Moreover, the sanitary state of military stations—particularly the cleanliness of living quarters, the disposal of waste, and the management of water and vegetation—varied both across different stations at any given time and across time at any given station.

Even in the worst epidemics and among soldiers of the same race who had been provided with precisely the same equipment and exposed to precisely the same conditions, some soldiers clearly fared worse than others. Physicians therefore paid close attention to the personal characteristics and behaviors of individual soldiers in search of factors that made them more likely to fall ill. In particular, they suspected that heavy food, strong drink, sexual promiscuity, and intense physical exertion were particularly risky for soldiers not yet acclimatized or "seasoned" to the tropics.[43] Physicians also paid close attention to the mental or emotional state of soldiers, noting that some were intensely fearful of falling ill while others remained sanguine about their prospects, and that the former seemed more likely to succumb to disease than the latter.[44] According to James M'Cabe, a medical officer stationed in the West Indies in the mid-1810s, this was partly because such a fearful state was damaging to health in itself, but also because those who succumbed to fear often indulged "in excesses which in any country or climate would ruin the constitution, until in the end they are really overtaken by the disease which their own imaginations had represented to them the impossibility of escaping."[45] "Moral" causes thus played a prominent role in M'Cabe's understanding of tropical disease, as it did in the theories of many other physicians of the time.[46]

Over the course of the mid-nineteenth century, intensified efforts by the British army to reduce disease rates also provided medical officers with new opportunities to observe relationships between the conditions of the tropics and the health of soldiers. From the 1820s onward, for example, the army's increasing reliance on hill stations such as Newcastle as refuges from tropical disease allowed physicians to observe the effects of moving a given individual or group of soldiers from one altitude to another.[47] The use of hill stations rested on a commonplace of early-nineteenth-century neo-Hippocratic medical geography—namely, that diseases such as yellow fever that were associated with the heat, moisture, and marshes common in coastal and lowland areas would be absent at higher altitudes and cooler temperatures.[48] The devastating yellow fever epidemic of 1840/41 led the commander of the British troops in Jamaica at the time, William M. Gomm, to note that British troops in lowland areas had "sunk under a malady from which the mountain stations of the height here adverted to are entirely exempt."[49] Gomm's decision to establish Newcastle as a refuge was a direct response to this observed difference in disease geography. Over the following decades, individual soldiers who sickened in the lowlands continued to be sent to Newcastle for recovery, while entire regiments were stationed in the hills during epidemics or seasons known to be especially hazardous or when a regiment of unacclimatized soldiers had just arrived in the tropics.

Only a minority of medical officers actively pursued investigations of the causes of tropical disease, but those who did increasingly had the chance to share their observations and theories with one another. As the number of British military stations in the tropics increased and the webs of transportation and communications that bound them together thickened over the course of the nineteenth century, medical officers found it easier to compare their own observations with those of others stationed throughout the empire. Even before their first posting to a tropical station, they could consult with more senior medical officers who had served in the tropics, and when they returned after years or decades of service, they passed on their own observations to those who were taking their place. Robert Armstrong, for example, who directed the British naval hospital in Jamaica in the 1820s before returning to England, decided to publish his manuscript on *The Influence of Climate and Other Agents on the Human Constitution* in part because he received more requests for advice from medical officers on their way to the West Indies than he had time to answer.[50] Once in the field, medical officers stayed in touch with their peers at other stations and in Britain through correspondence, travel, and other informal means. At the same time, the growth of medical publishing offered formal venues for British medical officers to share their observations, as did the issuance of annual reports by the Army Medical Department from the late 1850s onward.[51]

By examining the growth of these networks of medical officers, we can see how the bureaucratic and military structure of the empire provided some of the material and social conditions for the development of a new understanding of health. By the middle of the century, British medical officers had access to a large and growing store of observations on the role of external conditions and personal characteristics in producing disease in the tropics. They were part of a global network that not only spanned the British Empire but also connected them to the physicians of other nations, giving them unprecedented opportunity to map the distribution of diseases, compare the vulnerabilities of different types of people, and seek out correlations between particular conditions and the outbreak of epidemics such as the one that struck Jamaica in 1866/67. The movements of large numbers of soldiers from one part of the world to another—such as the movement of European soldiers to and from stations in the tropics—provided a series of unintended experiments that British medical officers were able to observe at close quarters. In all these ways, the bureaucratic, militarized, and tropical British Empire of the nineteenth century provided the infrastructure for the emergence of new understandings of health in terms of the relationships between individuals and their surroundings.

QUANTIFYING

In the mid-nineteenth century, researchers across a wide range of disciplines and areas of interest began to seek out masses of quantitative data in the hope of gaining new insights into the order of nature. The production and circulation of these data was made possible largely by the desire of the United Kingdom and other European states to better govern their territories and populations, including those of their far-flung colonial possessions. The availability of these data altered the practices of physicians and medical researchers, encouraging them to search for quantitative correlations among categories of people, the characteristics of their surroundings, and the prevalence of disease. The results of this search challenged the neo-Hippocratic idea that certain kinds of climates and landscapes were inherently unhealthy.[52] By providing evidence of "striking contradictions to the usual theories," as an 1838 report put it, statistical studies of the health of populations motivated a search for new explanations that accounted both for the nuances of local environments and for the varying characteristics of the people concerned.[53] Not described in explicitly environmental terms until the closing decades of the nineteenth century, these findings nonetheless laid the foundation for physicians of the era to view health and disease as products of relationships between organisms and their environments.

This statistical turn was part of a broader trend that linked science to government in new ways in the mid-nineteenth century. From the 1820s onward, rather than seeking to identify ideal or exemplary types of plants, animals, and other natural phenomena through the intensive study of unique specimens, researchers interested in subjects ranging from ocean currents to disease outbreaks to crime rates reframed their questions in terms of frequency, proportion, and averages that could be applied to masses of similar objects or individuals.[54] To characterize these complex and variable masses of data, they drew on techniques of quantitative comparison and calculation that collectively came to be known as *statistics*, a term that had been first used in the mid-eighteenth century to describe the study of matters relevant to states.[55] New institutions such as the Statistical Society of London, founded in 1834, provided forums for the sharing of such techniques across diverse domains and disciplines, including medicine. In part, it was these methods and their wide applicability that led the British polymath William Whewell to coin the term *scientist* in the 1830s, which he introduced to describe a new breed of specialists united by research methods and principles of explanation rather than by a common subject matter.[56]

The new "scientists" adopted these statistical methods as a way of making sense of massive quantities of new kinds of numerical data. The mid-nineteenth-century flood of such data arose in large part from the fact that imperial European states such as Britain were seeking to manage the health and productivity of their populations—a task that required not only detailed information but also new kinds of expertise on resources, trade, warfare, and disease.[57] With the help of Britain's expanding imperial infrastructure and the cooperation of researchers from other European nations, British scientists in the decades after Napoleon's defeat constructed data-gathering networks that went beyond even the most ambitious efforts of previous researchers. Drawing on the labor of hundreds or even thousands of observers, many of them serving in the military or the colonial bureaucracy, they were able to obtain precise, regular, standardized, and geographically distributed measurements that no single researcher—even someone as talented and widely traveled as Alexander von Humboldt, whose work inspired many of these efforts—could have amassed alone. In a study of ocean tides in the 1830s, for example, Whewell was able to both serve the needs of the British navy and commercial shipping and take advantage of the empire's expanding geographical reach.[58]

For British medical officers, the end of the Napoleonic Wars marked an important turning point in the adoption of these kinds of statistical methods and the construction of the observational networks they depended on. Many physicians had gained firsthand experience with the heavy military cost of disease during those wars, including the devastating effects of yellow fever in Haiti, and a few had begun collecting careful records on causes of death and illness among soldiers and sailors. Among them was James McGrigor, who had coordinated the collection of medical statistics as head of army medical services in the Peninsular War.[59] After McGrigor was appointed director general of the newly established Army Medical Department in 1815, he brought his empirical turn of mind to military medicine throughout the British Empire. One of his first initiatives was to require medical officers to submit standardized forms summarizing disease and mortality in the units they served on a semiannual basis.[60] Forced to record and to share their observations, physicians posted throughout the British Empire began to search for patterns in the facts that they were collecting and summarizing, and the Army Medical Department was soon in possession not only of a vast collection of medical data but also of a cohort of young, ambitious medical officers capable of analyzing it.

Compiled into what eventually amounted to more than one hundred and sixty volumes, the statistics that McGrigor had demanded nonetheless remained largely unused until the 1830s.[61] What brought them into use in that decade were new concerns about the cost of supporting veterans in Britain who had been invalided after contracting diseases in the tropics. In 1835, these concerns led the Army Medical Department to launch a statistical study of disease rates in the West Indies and throughout the British Empire. The effort was initially headed by Henry Marshall, a deputy inspector general of army hospitals, who had published a study of medical topography in 1821 based on observations he had made while stationed in Ceylon.[62] Dedicated to McGrigor, Marshall's study of Ceylon included the kinds of qualitative descriptions of topography, climate, and disease that were typical of the medical geography of the day as well as a number of numerical tables of disease rates based on the kinds of regular record-keeping that McGrigor had made mandatory a few years earlier. In 1835, partly because of this experience, Marshall was tasked with the much larger study of disease rates at military stations throughout the British Empire. He was aided by Alexander Tulloch, a junior officer with legal and statistical training who had begun collecting and analyzing data on soldiers' illnesses while stationed in Burma. After Marshall's retirement the following year, Tulloch carried on the project with the assistance of another British army physician, Thomas Graham Balfour.[63]

Centered on numerical tables extracted from the Army Medical Department's records, the reports published by Marshall, Balfour, and Tulloch in the late 1830s and early '40s characterized in quantitative terms the relationships among soldiers of particular types, the conditions in which they lived, and the prevalence of disease.[64] They featured abundant numerical tables that readers could use to quickly locate and compare rates of disease under intersecting conditions of various kinds. By displaying in neat rows and columns the number and proportion of cases of each kind of illness for a variety of categories—year or season, military station, civilian or soldier, enlisted man or officer, white or black, newly arrived from Britain or resident of many years, and so forth—such tables shifted attention away from the idiosyncrasies of individual cases and toward the regular relationships that could be detected in statistical masses. The authors of these reports used them to argue that disease arose neither solely from the characteristics of particular categories of people nor solely from the characteristics of the military stations where they served, but rather from the relationship between the two. In an excerpt on disease and mortality in the West Indies that Tulloch read before the Statistical Society in the summer of 1838, for example, such tables were used to show "how exceedingly variable is the mortality" from place to place, and specifically "how differently the climate of different islands affects the health of the troops."[65]

Despite having been required to submit semiannual returns of numerical data since McGrigor's reforms following the Napoleonic Wars, not all British medical officers warmed to the new statistical methods. The prominent role in the analysis and interpretation of these data played by Tulloch, who had no medical training or experience of his own, was one source of skepticism, as was the fact that a new kind of expertise, statistics, was being used to challenge medical truisms. Andrew Halliday, a deputy inspector general of army hospitals who had been stationed in the West Indies in the mid-1830s, was particularly scornful of the conclusions drawn from medical statistics and Tulloch's role in particular.[66] "A physician practically conversant with West

	Annual ratio of Mortality per 1000 of the White and Black Troops, serving at each of the following subordinate Stations.										Ratio of Mortality per 1000 of White Troops serving throughout the Island.
	Up-Park Camp.	Port Royal.	Fort Augusta.	Spanish Town.	Stoney Hill.	Port Antonio.	Falmouth.	Montego Bay.	Maroon Town.	Lucea.	
By Fevers	121·	93·9	55·5	141·1	70·5	126·	80·	150·7	15·3	63·2	101·9
Diseases of Lungs .	8·5	7·3	9·7	9·1	6·5	5·2	8.7	7·9	6·3	10·2	7·5
,, Liver .	·8	1·6	·3	·3	1·9	1·2	1·6	..	·3	·6	1·
,, { Stomach } { & Bowels }	5·8	5·9	3·3	4·8	5·8	6·9	2·6	5·6	3·7	4·5	5·1
,, Brain .	2·	2·4	1·3	3·7	1·2	2·8	4·8	10·2	1·3	1·9	2·6
Dropsies . . .	·8	·6	·7	1·6	1·9	2·4	2·6	1·1	2·4	1·9	1·2
All other Diseases .	1·7	1·4	2·7	1·8	2·4	4·8	2·3	3·4	3·4	2·6	2·
Total . . .	140·6	113·1	73·5	162·4	90·2	149·3	102·6	178·9	32·7	84·9	121·3

FIGURE 5. A comparison of mortality rates from several classes of disease at various British military stations in the Caribbean between 1817 and 1836, compiled and published by Alexander Tulloch in 1838. (Reprinted from table LX in Alexander M. Tulloch, *Statistical Report on the Sickness, Mortality, & Invaliding among the Troops in the West Indies* [London: W. Clowes and Sons, 1838], 70.)

India service, and at the same time a competent arithmetician," Halliday wrote acerbically in 1839, "might have brought into view many causes of sickness and mortality, that can never be discovered by any accumulation of figures, and to discuss many matters which I firmly believe could never be made the subject of mere calculation."[67] Statistical methods thus received a cold welcome from physicians committed both to neo-Hippocratic theories and to clinical experience as the only sound basis for challenging accepted wisdom.

Such skepticism prevented the analysis of army medical statistics from becoming a major endeavor well into the 1850s. Indeed, with official support at a minimum, it was kept alive during the intervening decades mainly through the persistence of Balfour and Tulloch. The key turning point came with the Crimean War of 1853–1856, the first major conflict among European powers since the end of the Napoleonic Wars, when Florence Nightingale, working in collaboration with Balfour, Tulloch, and other military statisticians and reformers, emerged as a powerful advocate for the systematic collection and analysis of medical statistics.[68] Having successfully brought public attention to the scandalously high mortality rates of British field hospitals during the Crimean War, she continued to advocate after the war's end for new sanitary measures to address "overcrowding, want of ventilation, want of drainage, imperfect water supply," and other factors contributing to soldiers' ill health.[69] Gradually extended to military stations throughout the British Empire, these reforms were

closely linked to the collection of statistical data, which Nightingale argued
would allow physicians to identify what she called the "removable causes" of
disease and to move quickly to ameliorate them.[70]

In 1859, Nightingale's advocacy of statistics and sanitation helped lead to
the establishment, under Balfour's direction, of a new Statistical Branch within
the Army Medical Department.[71] By the time the Army Medical Department
commissioned its investigation into the 1866/67 yellow fever epidemic in Jamaica, medical statistics was therefore a well-established tool for revealing the
connections between the conditions under which soldiers lived, their personal
characteristics, and the prevalence of disease. In fact, these methods had already been implemented on a smaller scale in relation to an outbreak of yellow
fever in Jamaica in 1856. In the aftermath of that epidemic, the head medical
officer in Jamaica at the time, Robert Lawson, had analyzed the prevalence of
the disease at various military stations and encampments in quantitative terms,
arguing they showed that neither the high altitude nor the milder climate of the
highlands was sufficient to protect soldiers from disease.[72] The commission on
the 1866/67 epidemic, influenced by Lawson's recommendations, went even
further in its use of statistical methods, circulating a standardized questionnaire
to physicians in every parish in Jamaica as well as to the commanding officers
of the army and navy. It then collected and synthesized the data from these
questionnaires and aggregated it with other sources of information, including
records of yellow fever cases kept by the principal medical officers in Jamaica
during the previous decades, as well as the analysis published by Tulloch
in 1838.[73] The result was both a case-by-case chronological narrative and a
comprehensive statistical view of the epidemic's impact on British troops in
Jamaica and the conditions that may have triggered it or hastened its spread.

In observing the slow adoption of statistical methods for studying military
health, from McGrigor's first record-keeping initiative in the late 1810s to the
establishment of the Statistical Branch a half-century later, we can see how a
set of techniques that were themselves not explicitly environmental could help
open the door to later environmental understandings of health by undermin-
ing the dominant theoretical frameworks of the time. These methods did more
than simply challenge some of the claims of neo-Hippocratic physicians, however. They also shifted the ground of the debate and the varieties of evidence
that were accepted as determinative. While the clinical experience cited by
Halliday did not lose its significance, it now had to compete with the kinds of
statistical observations advocated by Tulloch, whose lack of medical training
did not prevent him from challenging in bold terms the accepted wisdom of
the medical profession. Tulloch was himself hesitant to offer an alternative

theory; his work nonetheless helped open up a field in which others who were less hesitant sought to explain statistical variations in health in terms of disordered relationships between particular kinds of organized bodies and the external conditions they faced.

At the beginning of the nineteenth century, British physicians had no shortage of concepts and theories to explain the distinctive patterns of disease observed in the tropics. While in some ways these concepts and theories resembled the explicitly environmental ones that would emerge in the late nineteenth century, in other ways they were profoundly different. Influenced by neo-Hippocratic medicine and Enlightenment natural history, many physicians assumed that health was directly affected by the varying qualities of "airs, waters, and places." Moreover, they assumed that these effects were uniform across human types—or at least that any differences in susceptibility to the effects of climate could be moderated through acclimatization across lifetimes or generations. Since it was clear to any observer that the effects of climate on health varied across time, place, and individual, however, they also searched for other causes of imbalance in the relationship between the human body and its surroundings. These causes could be external or internal to the body; they included extremes of temperature and humidity as well as the specific weaknesses of an individual's constitution. Sickness arose, they believed, when the body became overstimulated or overheated, or when it suffered from deficits or excesses of blood and other fluids in certain organs or in the body as a whole. European soldiers in the tropics who indulged in immoderate consumption of food or drink or engaged in activities that exhausted their strength or exposed them to moisture, heat, or miasma, they argued, risked upsetting an already delicate balance.[74]

Throughout the nineteenth century, the precise reasons why some places in the tropics were less healthy than others remained the subject of vigorous debate among physicians adopting this neo-Hippocratic framework, in which the border between bodies and their surroundings was understood to be permeable rather than closed.[75] A place could be unhealthy, physicians believed, as could an individual, and under certain conditions the unhealthiness of the place could pass into the individual and vice versa. "Miasmas"—that is, "marsh poisons" or "soil exhalations" arising from putrefying organic matter— became a widely accepted explanation for the prevalence of fevers in certain locations at certain times of year.[76] John Hunter, who served as a medical officer in Jamaica in the early 1780s, attributed tropical fevers to "noxious exhalations from wet, low, and marshy grounds" that were produced by the combined effects of "heat, moisture, and decayed vegetable or animal matter."[77] These forces acted more powerfully in the tropics than in temperate climates, he argued, and newly arrived European soldiers were more vulnerable to them than those who had been acclimatized through long residence.[78] Proponents of this view believed that the heat and humidity of tropical climates somehow intensified the process of putrefaction or weakened the ability to Europeans to resist it, thereby producing the characteristic variations in disease prevalence from time to time and place to place within the tropics.

The rise of medical statistics from the 1820s onward challenged neo-Hippocratic theories about the connections between individual health and surrounding conditions, including climate and miasma. In the short term, its effect was mostly negative—that is, rather than offering a new conceptual framework for understanding health, it simply undermined the plausibility of neo-Hippocratic medicine as a whole. Sometimes the challenge was quite explicit. In 1839, for example, Tulloch argued that data on British troops in North America showed "the difficulty of establishing any uniform connexion [*sic*] between the presence of marshy ground, and the existence of those febrile diseases to which the exhalations from it are supposed to give rise."[79] More broadly, statistical studies showed that there were numerous cases where all the conditions for the production of disease through miasma seemed to have been satisfied but no disease occurred, and just as many cases where not a trace of miasma could be detected but disease nonetheless ran rampant. Still, Tulloch and his colleagues hesitated to make claims about what did cause tropical fevers, instead limiting themselves to challenging what they considered to be the premature generalizations of earlier medical geographers.[80] External conditions were obviously important to human health, they agreed, but none of the existing theories were capable of explaining how.

Not everyone was so reticent about proposing alternative models of disease causation. By the 1840s, some physicians were inspired by the results of the Army Medical Department's statistical studies to search for theories that offered more satisfying explanations of disease in the tropics. For this purpose, they turned to comparative anatomy and medical physiology.[81] Rather than claiming that bodies and their surroundings were mutually permeable, as neo-Hippocratic physicians had done, they argued that bodies could best be seen as functionally integrated entities that responded to external conditions in ways that were determined by their particular patterns of organization. In his 1843 book *The Influence of Climate and Other Agents on the Human Con-*

stitution, for example, Armstrong drew on French anatomy and physiology, British medical statistics, and his own experiments and firsthand observations in Jamaica to argue that miasmas—one of the supposed causes of tropical fevers—were nothing but "aerial and unsubstantial phantoms."[82] Diseases encountered in tropical climates were instead products of "the intimate relation which exists between external agents, the organization, and physical necessities of the individual."[83] From this perspective, no climate or place could be unhealthy in itself, and no disease could be transferred between a place and a human body. Rather, places or climates could only be unhealthy for certain kinds of organisms, and diseases could only be attributes of individuals, not of their surroundings.

Statistical studies also informed changing ideas about the medical significance of race. Drawing on studies showing continued vulnerability to tropical diseases even among Europeans who had lived in the tropics for many years, physicians challenged the Enlightenment idea that all varieties of humans were fundamentally the same, even if some had been changed in minor or major ways through long exposure to particular climates. On the contrary, they argued that each human race, like each species of plant or animal, was organized in a way that suited the specific conditions under which it lived, and that this form of organization could not be easily changed. Medical statistics provided seemingly incontrovertible evidence of the reality of these increasingly rigid racial distinctions. In the West Indies, for example, Tulloch's 1838 report showed that neither African nor European troops showed any signs of becoming more resistant to tropical diseases over time, suggesting that acclimatization, or "seasoning," was a myth.[84] On the contrary, the longer they spent in the Caribbean, the more likely they were to fall victim to diseases such as yellow fever.[85] Moreover, although both Africans and Europeans were more likely to fall ill in the Caribbean than natives, the specific diseases that afflicted them differed, suggesting that each race carried its own unique vulnerabilities.[86] Such analyses helped convince physicians that racial differences in disease vulnerability were real, permanent, and specific.

British medical statistics and French comparative anatomy and physiology thus constituted two of the pillars of the new race science of the mid-nineteenth century, which depicted human racial differences as specific, biological, and unchangeable. Tulloch himself explicitly drew such conclusions from his studies, arguing that "they point out the limits intended by Nature for particular races, and within which alone they can thrive and increase."[87] In Britain, the most infamous advocate of this view was Robert Knox, a surgeon who began his career as the Napoleonic Wars were winding down, spent several years as an army surgeon at the Cape Colony, and then studied with Étienne Geoffroy Saint-Hilaire, Henri Marie Ducrotay de Blainville, and others in Paris in the 1820s. After returning to Britain, Knox took up teaching and sought to apply their ideas about comparative anatomy and physiology to the study of human races.[88] In his 1850 book *The Races of Men: A Fragment*, Knox contended that each race was uniquely suited to the climate and other conditions prevailing where it was originally found and could only with great difficulty survive elsewhere. As evidence, he drew on Tulloch's studies showing that Europeans suffered more from disease in tropical countries than the races native to them—a finding that in Knox's view indicated the ultimate futility of European imperialism. "Withdraw from a tropical country the annual fresh influx of European blood," he wrote, "and in a century its European inhabitants cease to exist."[89] Controversial when published and subsequently rejected by most scientists and physicians, Knox's book nonetheless reflected the increasingly biological view of race in the mid-nineteenth-century and the use of medical statistics to support it.

Medical theorists who adopted biological perspectives during this period did not yet use the term *environment*, which would come into wide use in English only in the 1870s and '80s.[90] Nonetheless, they began to adjust the scope and connotation of other terms, particularly *climate*, in ways that brought them closer to the meanings that physicians at the end of the nineteenth century would express with *environment*. Knox, for example, believed that each human type was suited to a particular climate, but by *climate* he meant something much more specific than broad characteristics such as "tropical" or "temperate," or the kinds of variations in heat and humidity that had concerned eighteenth-century naturalists such as Buffon. Africans from tropical regions did not thrive everywhere in the tropics, he argued; nor could Europeans establish permanent colonies everywhere that temperate climates were found. Other factors were also critical, and it was the sum total of all these factors that determined the capacity of a particular "race" to flourish.[91] This expansion of the meaning of *climate* was even more explicit in the work of one of Knox's disciples in scientific racism, the anthropologist James Hunt, who argued that the study of "ethno-climatology" was the only path to a "correct and physiological system of colonization," particularly in the tropics.[92] Carefully distinguishing his use of the term *climate* from earlier uses, he noted that he meant it to include altitude, soil, light, water, wind, pressure, vegetation, diet, and indeed "the whole cosmic phenomena" capable of affecting the human organism.[93]

By examining the concepts deployed by mid-nineteenth-century physicians and race theorists, we can see a transition from an older, neo-Hippocratic way of understanding the body and its surroundings as mutually permeable, toward a biological way of understanding the body as a functionally integrated unit related in distinct ways to the external conditions it encountered. We can also see, however, that the idea that disease arose from a disordered relationship between an organism and its environment did not immediately sweep away the alternatives. Even as they adopted elements of the new biological perspective, physicians continued to explain disease and to practice medicine in neo-Hippocratic terms. This mix of old and new can be seen in the report on the Jamaican yellow fever epidemic of 1866/67, whose authors readily admitted the inadequacy of neo-Hippocratic theories concerning the effects on health of altitude, temperature, miasmas, acclimatization, and personal attributes, but who also continued to deploy those theories by speculating that the epidemic arose from some combination of a "general epidemic influence" or "epidemic constitution," local conditions, and the personal characteristics and behaviors of each individual.[94] Similarly, while overcrowding, uncleanliness, and intemperance among soldiers at Newcastle had not caused the outbreak, they argued, these factors did seem to have "assisted in causing the natural advantages of altitude and temperature in the tropics to be of less value than had previously been anticipated."[95] Rather than observing a sudden rupture between two radically different understandings of health, we can therefore see how mutually incompatible concepts and theories continued to coexist.

AN ENVIRONMENT OF GERMS

In the last decades of the nineteenth century, the idea that disease was the result of a disorder in the relationship between an organism and its environment received a powerful boost from microbiology, which demonstrated the existence of microscopic bacteria, parasites, and viruses capable of causing specific diseases.[96] In the case of yellow fever, the discovery of the disease agent and its mode of transmission gradually emerged between the 1880s and the first decade of the twentieth century. In contrast to their leading role in the study of yellow fever earlier in the century, British medical officers played a minor role in these developments, one mainly limited to providing methods and theories based on their research on tropical diseases elsewhere in the world. Their diminished role was linked to shifts in the structure of the British Empire, including the fading importance of the Caribbean in general and Jamaica in particular. As Britain's attention turned toward India, it was gradually supplanted as the dominant colonial power in the Caribbean by the United States, especially after the latter's occupation of Cuba and Puerto Rico as a result of the Spanish–American War of 1898. In this new imperial context, Cuban and US researchers took the lead in untangling the etiology of yellow fever. Their research helped show that tropical diseases could be understood as disorders in the functions of the human body arising from microscopic disease agents present in its environment.

Crucial to this new view of disease was the rise of the laboratory as a site of controlled observation and experimentation. Increasingly, biologists carried out their studies in enclosed spaces where every detail of the conditions of the organism or biological processes they were studying could be measured and manipulated. By the 1870s, techniques for growing colonies of bacteria and other microbes under artificial conditions and examining them under the microscope made it possible to identify the microbial agents that caused certain diseases. In 1876, for example, the German bacteriologist Robert Koch proved that anthrax was caused by a specific kind of soil bacteria.[97] Grown in the laboratory under the proper conditions, pathogenic bacteria could be injected into the "internal milieu" of humans or animals to produce disease, providing experimental proof of causation that statistical studies could not.[98] Among its other consequences, the identification of microbial disease agents transformed the understanding of the role of climate in causing tropical diseases. Important in this regard was the British physician Patrick Manson's discovery of the microbial cause of the disfiguring tropical disease filariasis in the late 1870s. Through microscopic investigations and experiments carried out during his long residence in East Asia, Manson found that the disease was caused by a parasitic roundworm transmitted by mosquitoes. This "tropical" disease, Manson showed, was not caused by the direct effects of tropical climate or miasmas on the human body, but rather by the transmission of a microscopic disease agent from one human to another via a third organism that served as what Manson called its "nurse," or intermediate host—namely, a specific kind of mosquito that happened to flourish under tropical conditions.[99]

Inspired by advances in bacteriology in general and Manson's discovery in particular, physicians reconsidered the etiology of yellow fever, a disease that had defied all attempts at explanation in terms of person-to-person contagion or the direct effects of climate, miasma, or other external conditions. In 1881, in a paper presented to the Royal Academy of Sciences of Havana, the Cuban physician Carlos Finlay argued that yellow fever was caused by an unknown disease agent transmitted through the bites of infected mosquitoes.[100] Although Finlay was not the first to suspect the involvement of mosquitoes in yellow fever, his paper challenged the accepted wisdom at the time as well as Finlay's own previous position—namely, that the unusually high prevalence of the disease in Cuba was due to the high alkalinity of its atmosphere. Instead, he now argued, the accumulating evidence of medical statistics and other observations rendered untenable "any theory which may attribute the origin or propagation of yellow fever to atmospheric, miasmatic or meteorological influences, or . . . to filth or neglected hygienic principles."[101] While capable of disproving neo-Hippocratic explanations of the disease, such observations did not definitely reveal the agent of yellow fever. For that, Finlay turned to the laboratory, and specifically to microscopic examinations of mosquito anatomy and experimental tests in which he exposed healthy subjects to mosquitoes that had previously bitten infected patients.[102] Although his initial findings were inconclusive (partly due to the difficulty of finding patients willing to be exposed to yellow fever), in subsequent years Finlay continued to advocate his theory of mosquito transmission and to search for the microscopic disease agent he was certain he would eventually find.

Over the course of the 1880s and '90s, even though physicians continued to view the specific disease vector and agents Finlay had proposed with skepticism, they increasingly accepted his basic premise: that the external conditions correlated with yellow fever outbreaks, such as tropical climates and marshes, should be examined not for their direct impact on human bodies but rather as conditions under which an organism capable of transmitting the disease agent, the "germ," from one human to another could flourish.[103] Among them was the US military physician and bacteriologist George Sternberg, who met Finlay in 1879 while visiting Havana as part of a yellow fever commission, the establishment of which reflected the United States' growing commercial and strategic interests in the Caribbean.[104] By the early 1890s, as a result of multiple visits to Cuba, ongoing conversations with Finlay, and his own experiments, Sternberg was fully convinced that "the specific infectious agent in yellow fever" was "a living micro-organism" that flourished under particular meteorological and sanitary conditions. Dubious of the evidence for any of the specific candidates that Finlay and others had proposed, he shared their certainty that the microorganism in question would eventually be found.[105]

The loss of thousands of US soldiers to yellow fever in the Spanish–American War proved to be a critical turning point. Following the war, Sternberg organized a commission under the command of US army medical officer Walter Reed to investigate the causes of the disease among soldiers at Columbia Barracks in Quemados, Cuba. Through studies of human subjects exposed to various hypothesized sources of yellow fever, the commission demonstrated that Finlay's hypothesis was largely correct—that is, that yellow fever was indeed transmitted exclusively by the species of mosquito he had identified, although the disease agent was later proven to be a virus rather than a bacterium.[106] Along with similar findings on malaria several years earlier by the British researcher Ronald Ross (which built on Manson's even earlier findings regarding filariasis), the discovery of the vector and agent of yellow fever provided proof that so-called "climatic diseases" were not the product of the direct action of tropical climates or miasmas. Rather, as Sternberg explained, such diseases "prevail only where climatic conditions are favorable for the propagation of the species of mosquitoes by which the parasites to which these diseases are due are transmitted from man to man."[107] In other words, the tropical climate itself posed little direct threat to the health of European soldiers. Its importance lay in the fact that it was one component of the broader set of conditions for the flourishing of pathogenic organisms that Sternberg called the "tropical environment."[108]

The proof that yellow fever and malaria were mosquito-borne diseases led to mosquito control and eradication efforts throughout the Caribbean in the early decades of the twentieth century.[109] In many respects, these efforts resembled older sanitary efforts that had been introduced before the development of bacteriology. Whether physicians blamed "marsh poisons" or marsh-dwelling mosquitoes for yellow fever, they were just as likely to recommend draining marshes located close to military stations or ensuring that stations were not located near marshes in the first place. Nonetheless, even if much remained the same, the new theories did lead to changes in sanitary practice. In addition to regarding sanitary interventions as ameliorating the conditions that were directly responsible for disease, sanitarians now saw such interventions as altering the environment of certain organisms in ways that made them less likely to transmit disease agents to humans. In some cases, this resulted in sanitary standards becoming more stringent. To eliminate disease-transmitting mosquitos, for example, it was often necessary to eliminate virtually all sources of standing water, including small containers that would have been ignored by an earlier generation of neo-Hippocratic sanitarians since they were not obvious sources of miasma. In other cases, standards became more lenient. Eradicators of the mosquito sought to cut grass and brush away from residences, for example; but unlike the sanitarians who preceded them, they only took action when vegetation was thick enough to harbor mosquito populations.[110] What mattered was not the vegetation itself but the habitat it provided for mosquitoes to thrive.

Precisely because the findings of germ theory were often compatible with the sorts of intervention they were already advocating on the basis of neo-Hippocratic theories of disease, sanitarians were among those who most enthusiastically embraced biological understandings of disease and the environmental language that came with them. The hygienist J. Lane Notter, for example, who served as a medical officer in Canada and Malta before training several generations of young physicians at the Army Medical School at Netley, was among the first British physicians to begin writing explicitly about "environments" in the 1890s.[111] In his contribution to the 1893 volume *Hygiene and Diseases of Warm Climates*, for example, he deployed the term *environment* alongside *climate*, giving each a distinct meaning.[112] For Notter, *climate* referred to sunlight, heat, humidity, and other characteristics of the atmosphere, while *environment* was the sum total of the surrounding conditions faced by the soldier, including conditions that were subject to human modification such as soils, drainage, wind, vegetation, sanitation, and exposure to pathogenic microorganisms in food and water.[113] In this sense, climate became a subordinate aspect of environment rather than the master concept of tropical medicine that it had previously been. In this new view, moving from a temperate to a tropical climate posed a threat to health not because certain climates were inherently unhealthy for European bodies but because a "change of environment" could lead to "deteriorations in the functions of the body."[114]

In addition to suggesting new kinds of sanitary interventions, this view of disease also reshaped the way physicians thought about human racial differences. Instead of focusing on the suitability of different human races for tropical climates as such, as had mid-nineteenth-century race theorists such as Knox and Hunt, physicians turned their attention to immunity or resistance to specific disease agents. They argued, for example, that people of African descent had anatomical or physiological traits that made their bodies inhospitable environments for disease agents common in the tropics or resistant to the living vectors that transmitted them. In his original 1881 article on mosquitos as vectors of yellow fever, for example, Finlay reformulated older ideas of racial difference in terms of the new biology by speculating that qualities of the skin or blood of people of African descent might have chemical or physical properties that made them resistant to the transmission of yellow fever by mosquito.[115] Even those who rejected the existence of these kinds of specific biological differences in disease vulnerability continued to use tropical medicine to support claims for deep-seated racial differences. After mosquito eradication and other interventions made it possible to reduce the threat of infection for white visitors to the tropics, race theorists argued that Europeans' racially specific ingenuity had allowed them to flourish even in environments to which they were biologically ill suited, making them perhaps the one race capable of colonizing the entire Earth. Rather than entirely undoing theories of racial difference based on differential disease mortality in the tropics, germ theory helped to shift their focus from biological differences in adaptation to specific environments to biological differences in the capacity to technologically adapt to any environment whatsoever.[116]

We can therefore see that when, in 1902, Sternberg wrote about the "tropical environment" in which yellow fever and the mosquitoes that transmitted it both flourished, the language he was using was still relatively new for physicians, even if talk of "environments" had been growing steadily across a variety of fields since the 1850s. Moreover, we can see how the explicit use of the term *environment* did not emerge in the context of neo-Hippocratic medicine, with its ideas of the balance of humors, the interpermeability of airs, waters, and bodies, and the inherent healthiness or unhealthiness of particular places and climates, even though many of those ideas now seem to us essentially "environmental" in nature.[117] Instead, the explicit adoption of the concept of environment first emerged in the context of efforts to complement or even replace theories of the direct influence of "airs, waters, and places" with a biological understanding of tropical disease as the result of pathogenic microbes and their living vectors flourishing under tropical conditions.[118] As Sternberg, Notter, and others argued, these microbes and vectors constituted an important part of the "environment," a new object of study and concern that was distinct from the neo-Hippocratic understanding of the influence of the climate and other "surrounding things."[119] What one historian has described as the "germ-theory theory of environment" was therefore the first explicitly environmental theory of health.[120]

*

In its instructions to the commission investigating the 1866/67 yellow fever epidemic in Jamaica, the Army Medical Department had stressed the importance of gathering detailed information and analyzing it "with the same rigorous exactness that would be employed in a chemical or physical investigation."[121] Like the geological and meteorological surveys that were taking place around the same time and whose results it also drew on, the commission's survey depended on the existence of large populations of objects that could be treated as if they were chemical or physical substances—that is, as collections of essentially identical objects that might behave in unpredictable ways when considered individually but which obeyed hidden laws that could be identified when they were studied as a mass. In the case of military medicine, the masses in question consisted of soldiers and sailors, and the observations consisted of medical reports detailing the prevalence of disease under various conditions. The demand for exactness reflected decades of statistical analyses showing that the health of broad classes of human beings could be characterized in quantitative terms; it also reflected physicians' increasing adoption of concepts and techniques from comparative anatomy and physiology, which viewed the human body as a functionally integrated organism whose character could only be understood in relation to its surrounding conditions. Medical statistics thus played a key role in the emergence of an "environmental" perspective on health in the second half of the nineteenth century.

Examining the adoption of the concept of environment in British medicine in the nineteenth century helps us see what made it distinct from the understandings of health that preceded it and why physicians in the last decades of the nineteenth century felt the need to adopt a new term to describe it. Paradoxically, the emergence of an explicitly environmental view of health depended on the rejection of longstanding neo-Hippocratic theories that had focused physicians' attention on climate, miasma, and topography and on the transfer of humors and diseases between places and bodies—all of which in retrospect seem eminently environmental. What led sanitarians in the late nineteenth century to adopt the concept of environment was not the triumph of neo-Hippocratic medicine, however, but rather its displacement by a new biological understanding of health according to which disease resulted from disordered relationships among the organism's parts or between the organism and its surroundings. The growth of microbiological accounts of disease from the 1870s onward—including the work of Finlay, Sternberg, Reed, and others on yellow fever—advanced this biological perspective even further. It showed that European soldiers sickened in the tropics not because of the direct effects of heat or humidity but because the tropics provided a suitable environment for disease-bearing organisms, particularly the mosquitoes that carried the microscopic agents responsible for yellow fever and malaria. Far from doing away with environmental explanations of disease, germ theory was thus responsible for making "environment" a key concept in medicine.

The emergence of an explicitly environmental perspective on health among British physicians was closely linked to the nineteenth-century expansion and reorientation of the British Empire toward the tropics, as well as to the racial politics that accompanied it. The globe-spanning infrastructure of the empire and the mobility of people, goods, diseases, and information that it enabled gave physicians new opportunities to observe the impact of changes in climate, topography, sanitation, and other external conditions on humans of different types. Their conclusions, in turn, were used to support new theories of biological difference among human races. These theories were centered on the idea that each race was characterized by a distinct pattern of biological organization that determined its suitability to particular environmental conditions. Unlike the neo-Hippocratic theorists who preceded them, advocates of these theories did not associate races solely with broad climatic regions such as "the tropics," nor did they think that rapid acclimatization was possible. Rather, they identified sets of external conditions for which they believed people of each race were specifically and permanently suited. In the British Empire, the adoption of "environment" as a medical concept thus accompanied and reinforced the biologizing of human racial difference and the increasingly rigid racial hierarchy that it was used to justify.

When late-nineteenth-century British physicians adopted "environment" as a key concept for understanding human health, they also sought to redefine the borders of human solidarity. Since they believed Europeans were biologically ill suited to tropical conditions, they saw projects of sanitation or mosquito control as aiming primarily to improve the environment of Europeans displaced to the tropics rather than all humans in the tropics. Even when they took into consideration the health of people of African descent in places like Jamaica, they assumed that their "environments"—that is, the external conditions that were specifically relevant to their health in light of how their bodies were biologically organized—were distinct from those of Europeans. Thus, rather than using the concept of environment to bring people together in a common project, these physicians deployed it to show that what might at first glance appear to be a world in common was in fact a diversity of worlds, each encountered and experienced in distinct ways by people of different races. This was not the only way that the concept of "environment" could be made to matter, however. The next chapter turns to urban reformers in the United States around the turn of the twentieth century who adapted the concept of "environment" to address social disorder rather than sickness, while also trying, with uneven success, to forge solidarity across human groups rather than deepening the differences between them.

CHAPTER 3

The Urban Milieu:
Evolutionary Theory and Social
Reform in Progressive Chicago

Set back from the street and surrounded on three sides by an elegant veranda supported by Corinthian columns, Hull House was built in 1856 in grand Italianate style in the midst of what was then a fashionable district of Chicago.[1] By late 1880s, however, both the house and the neighborhood had fallen on hard times. The house's owner, a real estate developer named Charles J. Hull, had long since moved elsewhere, and the house had been rented out for use first as an old-age home, then as a secondhand furniture store, and finally as offices and storerooms for an adjacent factory.[2] The area around it also had changed. Located on the city's West Side, halfway between the great stockyards to the south and the shipyards on the Chicago River to the north, it was now surrounded by communities that were diverse in national origin and socioeconomic status. Recently arrived Italians, Germans, Bohemians, Poles, Russians, and French Canadians lived alongside longer-established English-speaking families, while swaths of extreme poverty were interrupted here and there by small pockets of working-class prosperity. On the whole, the area was chaotic, unsanitary, crime-ridden, badly governed, and socially divided. As Jane Addams described it in the early 1890s, "The streets are inexpressibly dirty, the number of schools inadequate, factory legislation unenforced, the street-lighting bad, the paving miserable and altogether lacking in the alleys and smaller streets, and the stables defy all laws of sanitation."[3]

Addams came to know Hull House and the area around it as the result of a project of social reform that she launched in 1889 with a companion, Ellen Gates Starr. Inheriting significant wealth from her father, a prominent Illinois businessmen and politician, she had spent much of the 1880s searching for a way to put her social ideals into practice.[4] After a visit to East London's Toynbee Hall in 1888, she and Starr resolved to create a similar "settlement house" in Chicago, where middle-class reformers such as themselves would provide social services to the mostly working-class immigrant population living nearby. One of the first of hundreds of settlement houses established in the United States between the 1880s and 1920s, Hull House helped residents navigate the municipal bureaucracy, provided rooms for club meetings and musical rehearsals, hosted lecture courses and reading groups, operated a nursery and kindergarten, and sought to improve garbage collection and street maintenance in the surrounding districts. An extraordinary success almost from the moment of its founding, Hull House continued to expand over the subsequent decades under Addams's leadership. It acquired and constructed new buildings, established partnerships with other civic organizations and government agencies, and engaged with an ever-greater range of urban issues, from public playgrounds to political corruption. Self-consciously distinct from charity organizations that focused on direct aid to individuals deemed morally deserving, it became the leading example of a reform movement that instead sought to improve the welfare of individuals and communities by reshaping their surroundings.

The settlement movement of which Hull House was a leading example also pioneered new methods of social research, including a collection of methods that came to be known as the social survey. At a time when sociology was still establishing itself as a distinct academic discipline, settlement house residents crafted their own distinct approach to studying society, bridging practical concerns of social reform with general theories of social progress. They surveyed the social characteristics of the neighborhoods around them, plotting the results on maps and arraying them in statistical tables to visualize the impact of external conditions on the organization of society. At the same time, they learned from their everyday engagement with the people among whom they lived and from the successes and failures of their practical reform efforts, which served as tests of the power of environmental reform to transform the city's social fabric and to improve the welfare of its inhabitants and of the "social organism" to which they belonged.[5] The industrial city was central to their project, serving simultaneously as a site of reform, research, and residence. Chicago in particular seemed to epitomize not only the astonishing pace with which industrialization, urbanization, and immigration were transforming the United States but also the unprecedented social problems that came with them. It was thus an ideal site for research on the relationship between the material conditions of homes, streets, schools, and factories and the health and well-being of the city's inhabitants.

By the time the settlement movement was launched in the 1880s, the concept of environment was already being adopted across a wide range of fields, and well-read settlement house residents such as Addams were familiar with the definitions of the term offered in the mid-nineteenth-century by theorists such as Auguste Comte and Herbert Spencer. Nonetheless, such reformers did not simply accept the definitions that were offered to them or the tools and practices through which those definitions had come to matter under other circumstances. Instead, they crafted a distinctive understanding of the relationships between humans and their surroundings that centered on the idea that social evolution depended on humanity's intentional efforts to improve its relationship with its surroundings. True progress, they argued, would emerge only from efforts to improve the human condition that emerged organically and democratically from collaboration between working-class immigrants and middle-class reformers. Their work helped introduce the idea of the "social organism" and "social environment" into wide circulation. Indeed, they were so successful in doing so that Chicago-based sociologists in the 1910s and '20s increasingly argued that the social environment could be studied as a distinct realm of its own, one that was effectively independent of the material environment on which the settlement movement had focused much of its attention.

URBANIZING

When the World's Columbian Exposition opened on Chicago's South Side in 1893, its monumental neoclassical architecture, orderly and well-lit boulevards, and spectacular displays of electric light and other symbols of American innovation and prosperity entranced Chicagoans and visitors alike.[6] In the aftermath of the exposition, however, as the financial Panic of 1893 sent the United States into a deep recession, the residents of Chicago were confronted once again by the harsh realities of the industrial city, exacerbated not only by the economic crisis but also by the loss of the jobs that had been created by the exposition. According to one estimate, 100,000 men were thrown out of work virtually overnight.[7] Dependent as it was on low-skilled and industrial employment, the neighborhood around Hull House was hit particularly hard.[8] Even though it had mostly survived the destruction of the Great Chicago Fire of 1871, it had been profoundly reshaped by the growth of stockyards, shipyards, and other industries over the subsequent two decades, as well as by an influx of immigrants seeking to take advantage of the city's opportunities. Its workers spent their days immersed in parts of the industrial economy that were just as spectacular as the world's fair but much harder to envision as part of a utopian future, and at the end of the workday they returned to homes that had yet to see most if any of the modern amenities it had displayed.

For social reformers, the rapid growth and industrialization of Chicago in the late nineteenth century provided opportunities to observe the effects of the urban environment on people of widely varying backgrounds, occupations, and habits and on the social groups of which they were a part. In the decades following the Civil War, the westward expansion and economic development of the United States had accelerated, resulting in the explosive growth of new cities such as Chicago and the rapid transformation of older cities such as New York, which became sources of capital for industry in the US West and destinations for the goods and materials it produced. Chicago in particular grew from a small frontier outpost into the dominant metropolis of the Midwest. With the help of transcontinental railroad networks and the cheap and abundant coal that fueled them, as well as its location at the mouth of the Chicago River on the southwestern shore of Lake Michigan, it was transformed seemingly overnight into a bustling center for collecting, storing, sorting, processing, and pricing the products of the Great Lakes and Great Plains regions. Lumber, wheat, cattle, pigs, and other commodities flowed into the city and then were shipped onward to consumers in the densely populated eastern states.[9]

So rapid was Chicago's transformation in the second half of the nineteenth century that the city itself became renowned as spectacle of second nature, a world remade by human hands and impersonal economic forces.[10] Over the course of the late nineteenth century, even as the ties that bound it to its hinterland were tightened, the links connecting the city to its surroundings became increasingly difficult for the average resident to perceive. The stockyards and slaughterhouses on the South Side and the grain elevators that stored wheat and other grains for eventual transport to the East were so gigantic in scale and industrial in character that even though it was obvious that they had to be continuously fed by rural ranches and farms outside the city, it was easy to see them as entirely urban phenomena. Similarly, the hardworking horses who drew carts, carriages, and buses reminded observers less of their countryside counterparts than of living machines harnessed to the city's relentless pace.[11] Over time, even these hints of Chicago's connection to the countryside became increasingly obscure as stockyards and slaughterhouses were pushed to the margins and horses were gradually displaced by electric trolleys and eventually by the internal combustion engine. When "nature" was reintroduced through the creation of urban parks, such as the new green spaces designed for the 1893 world's fair by the landscape architect Frederick Law Olmsted Jr., it was not a "complete illusion" of nature that was being presented to viewers, as one attendee claimed, but rather a highly selective one.[12] Designed for the comfort and leisure of the city's inhabitants, such parks resembled English landscape parks or the rural sites where wealthy Chicagoans spent their weekends and summer holidays more than the working farms and forests that supplied the city's needs.[13]

The city's explosive growth in the late nineteenth century was also accompanied by a marked change in the character of its human population. With the seemingly insatiable demand for labor in its factories, stockyards, shipyards, railyards, and docks, Chicago attracted many of the immigrants who flooded into the United States in the decades following the Civil War. A large proportion of them hailed from eastern and southern Europe, but they also included internal migrants, including a small number of African Americans escaping the limited opportunities and racialized violence of the post-Reconstruction South. These migrants were both pushed and pulled toward Chicago. Displaced from their homelands by poverty, war, oppression, and economic forces that were transforming agriculture along with industry, they chose Chicago and other industrial cities of the United States because of the economic and social opportunities they promised. Once in these cities, they found themselves living next to people from a variety of ethnic and religious backgrounds under unfamiliar and often challenging conditions. Having never lived in a large town before, let alone a city of Chicago's magnitude and level of industrial development, many of them struggled to survive.

Even native-born Chicagoans struggled in the face of its exploitative industries, inadequate infrastructures, and faltering political institutions. These collective failures made for a hotbed of conflict between social classes, which sometimes erupted into violence, as it did in the notorious Haymarket Riot of 1886, when multiple police and labor activists died after violence broke out at a demonstration for workers' rights. While such dramatic events sparked interest in reform, just as troubling to many social reformers were the subtler signs of a gradual breakdown in the standards of civilized society.[14] The crowded tenements, crumbling houses, and inadequate sanitation of Chicago's most impoverished districts regularly led to outbreaks of diseases such as tuberculosis and typhoid; meanwhile, the proliferation of crime, prostitution, drunkenness, gambling, poverty, homelessness, and illiteracy attested to the growth of various forms of social and moral disorder. Cities had long been stigmatized as unhealthy and immoral, but the idea that they were growing even more so over time in one of the world's richest and most advanced nations challenged the prevailing late-nineteenth-century faith in progress, as did the failure of a corrupt and incompetent municipal government to meaningfully address the problem.

Taken together, these rapid changes in the urban landscape made Chicago and other industrializing US cities seem not only like a unique and unprecedented environment for human life but also one in desperate need of reform. Although the United States was not alone among industrializing nations in facing such problems during this era, its response to these changes was distinctive. Due to a tradition of decentralized government, powerful industrial cartels, and private philanthropy, government was expected to play a limited role in solving social problems. To the extent that the United States had anything like the social policies that had been put in place over the preceding decades in other industrialized countries such as the United Kingdom and Germany, the benefits of those policies were limited mainly to people deemed to be morally deserving, such as war veterans or widowed mothers.[15] In the absence of a strong state, Christian charity organizations took up some of the task of supporting the poor, but their interventions were typically focused more on the salvation of individual souls than on the structural conditions that shaped those individuals' lives and choices. Both the government initiatives, such as they were, and the Christian charity organizations distinguished between those who suffered through no fault of their own and those who allegedly deserved their fate. In this context, the settlement movement that arose in the 1890s and 1890s offered an approach that was distinctively different, even radical—one that combined sympathy for the unfortunate with an understanding of social problems as the result not of individual moral failings but of a disordered relationship between communities and the industrial conditions surrounding them.

In the emergence of a social reform movement focused on the conditions of the industrial city, we can see how the concept of environment was adapted to circumstances and aims for which the concept of nature seemed largely beside the point. At the same time that conservationists and preservationists such as the forester Gifford Pinchot, the trophy hunter (and US president) Theodore Roosevelt, and the wilderness advocate and Sierra Club founder John Muir were battling over the fate of what they understood as the vanishing US frontier, late-nineteenth-century urban reformers such as Addams, Florence Kelley, and Alice Hamilton were focusing on the material conditions necessary for social progress in an industrial society.[16] It was among the latter, not among the conservationists and preservationists of "nature," that the concept of environment first became central. Inasmuch as both of these groups influenced later "environmentalists," they can legitimately be seen as constituting two of the many "roots" and "origins" of environmentalism. In their own time, however, only one group was properly and explicitly "environmental"—namely, the one that conceived of the challenges of urban life as part of a much longer evolutionary narrative in which organisms both adapted to and reshaped their surroundings.

SURVEYING

While some settlement houses focused exclusively on providing social services and advocating for practical reform, others combined such activities with research aimed at understanding the causes and character of social problems. At Hull House in particular, Addams and other residents pursued an active program of research that took advantage of their proximity to diverse communities of immigrants struggling to thrive under the conditions of one of the late-nineteenth century's most dynamic industrial cities. They drew inspiration in part from a monumental survey of London's poor and working-class populations led by Charles Booth, which had used interviews and direct inspection of the living conditions of the poor to understand the origins and consequences of social problems in the city's East End (where Addams was shocked by the evidence of "hideous human need and suffering" during a visit in 1883).[17] Booth's study provided a model for how the causes and consequences of poverty could be rigorously studied in Chicago and other American cities.[18] Addams and other settlement workers were also inspired by the nascent science of "sociology," a term introduced by Auguste Comte in 1839 to describe "the positive study of the ensemble of fundamental laws concerning social phenomena."[19] By the end of the nineteenth century, sociology had begun to coalesce as a distinctive discipline with its own departments, journals, methodological standards, and fundamental questions. Drawing these influences together, the US settlement movement developed its own distinctive approach to the social survey as a way of producing knowledge about the relationship between social disorder and the material environment.

In its prototypical late-nineteenth-century form, the social survey involved a mix of observation and interviews, both of which were guided by the use of preprinted forms (or "schedules") that ensured that each surveyor (or "enumerator") gathered standardized data that could be easily mapped and tabulated.[20] By canvassing a neighborhood and plotting the resulting data on occupation, income, national origin, and housing type, social surveyors linked ethnicity and socioeconomic status to the spatial arrangement and physical infrastructure of the city in quantitative terms.[21] Hull House was responsible for organizing the first such survey in the United States. In 1893, under the direction of the newly arrived resident Florence Kelley and with funding from the US Department of Labor, it sent a team of surveyors throughout the neighborhood east of the settlement house to take notes on the conditions of the buildings and apartments and to ask residents about their nationalities, their families, their employment, and their wages. *Hull-House Maps and Papers*, published the following year, included color-coded maps of the distribution of populations with various demographic characteristics across the neighborhood, along with essays by Hull House residents detailing the neighborhood's distinguishing features and problems.[22] Further surveys were carried out in subsequent years by Hull House and other settlement houses throughout the United States. In the late 1890s, for example, with *Hull-House Maps and Papers* as his model and with the support of the University of Pennsylvania and the College Settlement House, W. E. B. Du Bois carried out a social survey of Philadelphia's largely African American Seventh Ward, the results of which were published in 1899 as *The Philadelphia Negro*.[23] Probably the most ambitious of these social surveys was the Pittsburgh Survey, conducted in 1907/08 in collaboration with the city's Kingsley House settlement, which offered a comprehensive account of the working and living conditions of the city's working class.[24] Like the social surveys in Chicago and Philadelphia, the Pittsburgh Survey sought to shift attention away from the moral failings of individual employees, employers, tenants, and landlords and toward the environmental causes of social problems.[25]

As the settlement movement grew and diversified, the social survey was refined and extended in new directions. In addition to mapping and tabulating neighborhood distributions of income and ethnicity, for example, surveyors began to conduct intensive investigations of the interiors of homes and workplaces. In the summer of 1900, with the sponsorship of Chicago's City Homes Association and authorization from the city's Department of Health to inspect buildings with or without the acquiescence of their owners, a team of enumerators conducted a detailed survey of the intimate domestic spaces of poor and working-class Chicagoans living in the city's tenements.[26] Combined with existing insurance and real estate maps, the survey's results showed how the health and morality of the city's working population and the prosperity of the city as a whole were threatened by the shortsighted greed of landlords and the failures of municipal government. Indeed, rather than improving with time, the quality of housing for the city's workers seemed to be steadily declining

FIGURE 6. A child in an alleyway in one of Chicago's tenement districts around 1900. (Reprinted from Robert Hunter, *Tenement Conditions in Chicago: Report by the Investigating Committee of the City Homes Association* [Chicago: City Homes Association, 1901], 40.)

as landlords neglected maintenance and crowded their lots with as many tenements and paying tenants as possible. Robert Hunter's report on the survey, *Tenement Conditions in Chicago*, singled out for special opprobrium the rear tenements that opened directly onto the stables, privies, and refuse piles of the alleyways.[27] "If landlords, with greed for profits and economy of ground space, continue to erect such tenements," he warned, "the city man will soon have new conditions to confront. The factory by day, the tenements by night, will be his environment."[28]

Social surveyors also investigated workplace environments, sending inspectors into factories to assess conditions and interview workers and conducting detailed comparative studies of the causes of occupational injury and disease. Informed by the US labor movement and European socialism but keeping both at arm's length, they reframed workplace injuries and illnesses as products of the environment rather than as the result of workers' inexperience, carelessness, or stupidity. Kelley, for example, who had translated Friedrich Engels's *Condition of the Working Class in England* into English before joining Hull House, focused her attention on the dangerous conditions faced by children laboring in Chicago's factories, eventually helping to enact significant reforms to Illinois's labor laws.[29] For young children, she successfully argued, no factory was a safe and healthy environment, no matter how well managed it might be. Alice Hamilton, a physician who took up residence at Hull House in 1897, began her work by investigating health and sanitation among Chicago's poor before turning her attention to workplace health and safety. In 1910, she was recruited to participate in an occupational disease commission for the state of Illinois and then to the US Bureau of Labor, where she investigated cases of toxic poisoning among workers in the white-lead industry and other dangerous trades.[30] Lax record-keeping, high rates of worker turnover, and the recalcitrance of all but a few progressive factory owners made this difficult work, but Hamilton eventually obtained compelling evidence that some factories were more dangerous than others, even within the same industry, the same city, and the same population of workers.[31]

Since the ultimate aim of social surveyors was social reform, it was essential for them to communicate their findings in ways that would sway politicians, philanthropists, and the broader public. To do so, they developed strategies of textual and visual representation that illustrated the importance of the physical and social conditions of the industrial city in shaping the lives of individuals and in producing social disorder. Rather than focusing on the moral qualities of individuals (as the literature of the Christian charity movement had often done), they depicted the city's residents as being immersed in and often overwhelmed by their surroundings. Recognizing the rhetorical limits of scientific prose and statistical tables, reports such as *Tenement Conditions in Chicago* and the Pittsburgh Survey's *Homestead: The Households of a Mill Town* also deployed photography and personal narratives. They included images of impoverished immigrant children playing amid the garbage of the alleyways or along the banks of sewage-filled streams as well as stories of individual suffering, such as the family of seven forced to share a two-room, one-bed apartment with a view onto a trash-filled alleyway, or the woman whose eyes filled with tears when she spoke of keeping her children indoors because she was afraid of exposing them to the filth and danger of the streets.[32] Beyond the social survey, an emphasis on the struggle of the individual against a constraining environment became one of the hallmarks of photographic and literary naturalism during this period.[33] Novels such as Theodore Dreiser's *Sister Carrie* and Upton Sinclair's *The Jungle*, both of which feature protagonists who migrate to Chicago from rural areas in search of a better life, revealed the capacity of urban conditions to reshape the lives of the poor and the powerless.[34]

Developed at a time when sociology as an academic discipline was still in its infancy, the social survey provided methods and data that US sociologists adopted not only to empirically test their theories, but also to prove the existence of something called "society" that could not be reduced to either

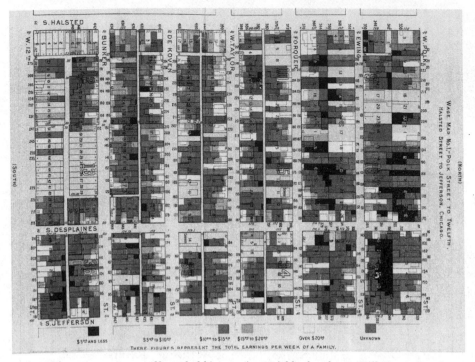

FIGURE 7. A map of household income in a neighborhood near Hull House, based on a social survey organized by Florence Kelley in 1893. (Reprinted from Residents of Hull-House, *Hull-House Maps and Papers* [New York: Thomas Y. Crowell, 1895]; in color in the original. Digital scan courtesy of Leigh Bienen, Northwestern University Library, and Northwestern Pritzker School of Law.)

biology or history and that required the use of a distinctive set of "sociological" methods. In general, sociology as an academic discipline in the 1890s remained loosely defined and methodological eclectic, drawing on methods and insights from history, economics, biology, and psychology. The University of Chicago's sociology department had been established in 1893, just three years after Hull House and one year after the new university itself, under the direction of Albion Small. Through his frequent contact with Addams, Kelley, and the other residents of Hull House, Small became convinced that the social survey would become a vital source of evidence for sociology. In the 1894 textbook *An Introduction to the Study of Society*, he and his coauthor George Vincent noted that one of the research methods they were advocating was exemplified by "series of sociological maps soon to be published under the supervision of Miss Jane Addams, of *Hull House*, in Chicago."[35] For their part, settlement residents consulted with academic sociologists to refine their methods and used publications such as Small's *American Journal of Sociology* to bring their work to new audiences.[36]

Through the history of the social survey and its central role in the settlement movement, we can see how an initially quite abstract and general concept became an effective tool for understanding and changing the world under a particular set of conditions. As theoretically alluring as the concept of the "social environment" may have been, by itself it was too ethereal to offer much purchase for practical reform. The consequences of disorder in the relationship between the social organism and its environment were visible in the poverty, disease, filth, crime, and vice of Chicago and other industrial cities, but the relationship itself remained difficult for reformers to grasp—even when, like Addams, they took up residence in the most troubled urban districts. It was only through studies such as *Hull-House Maps and Papers*, *The Philadelphia Negro*, *Tenement Conditions in Chicago*, and the six volumes of the Pittsburgh Survey that the "social environment" became a compelling object of both scientific research and social reform. We can thus see how the social survey transformed the social environment into a concrete object of knowledge and concern by statistically characterizing the social and material aspects of entire districts and their human populations.

REFORMING

Rather than seeking to exhort, cajole, or coerce poor and working-class immigrants into improving themselves through sheer force of will or religious devotion, the settlement movement sought to transform their surroundings while also giving them the skills they needed to transform them on their own. Its most powerful tool for doing so was the settlement house itself, which provided a working and living environment for residents as well as a gathering place and model home for people in the surrounding neighborhoods. At Hull House, for example, Addams and Starr took pains to decorate its public spaces in the style of a middle-class household rather than that of a hospital, clinic, school, or other institutional space.[37] Their aim in doing so was twofold. First, they sought to make it clear to visitors from the neighborhood that Hull House was their home as well as their place of work and therefore that they were not only trying to fix the neighborhood's problems but were also living with them on a daily basis. Second, they aimed to provide a model of middle-class standards that members of the community could enjoy during visits and to which they might aspire for their own homes. By making Hull House's rooms available for family celebrations and club meetings, for instance, they gave working-class and immigrant families an alternative to the saloon while exposing them to middle-class standards of cleanliness, wholesomeness, and morality. Similarly, Hull House's nursery and kindergarten provided not only a service to the community but also a model that settlement house residents hoped community members would emulate.[38] Courses in cooking, child-rearing, and sewing and other crafts made some of these lessons explicit.[39]

The settlement house also served as an educational institution in a broader sense, teaching neighborhood residents practical skills and giving them an opportunity to learn about and debate the issues of the day. Discussion clubs on politics, economics, and literature met at the house, and lecturers from the University of Chicago were invited to teach extension courses. Wary of alienating immigrants and creating intergenerational schisms by demanding that they abandon their traditions, Hull House also sought to incorporate into its educational program some of the immigrants' own practices. As Addams argued, although those practices were often rooted in rural landscapes and ways of life that had little in common with late-nineteenth-century Chicago, they nonetheless represented important stages in the evolution of human industry. To illustrate the point, Hull House residents assembled a small labor museum from a collection of spinning and weaving tools donated by Syrian, Greek, Italian, Russian, and Irish women in the neighborhood. By revealing a history of gradual progress from primitive hand spindles to modern weaving machines, Addams wrote, the museum enabled "even the most casual observer to see that there is no break in orderly evolution if we look at history from the industrial standpoint."[40] In such ways, Hull House sought to provide an educational environment in which components of the unfamiliar and often chaotic city were arranged so as to reveal a hidden order.

Recognizing that the impact of the settlement house would be limited as long as the surrounding neighborhood was hazardous to the safety, well-being, and moral character of its inhabitants, residents also advocated for improvements to public infrastructures and municipal services beyond the settlement house itself. Much of this work was informal, seeking to make up for official neglect and community apathy by persistently bringing problems to the attention of the municipal government.[41] In a few cases, however, it involved taking official positions as inspectors who were legally empowered by municipal, state, or federal governments to demand improvements. Addams, for example, briefly served the city of Chicago as a refuse inspector for the district around Hull House, following garbage collectors to ensure that they fulfilled their responsibilities.[42] Similarly, Kelley used her position as factory inspector for the state of Illinois to advocate for restrictions on child labor, while Hamilton's service on a state commission on occupational diseases allowed her to advocate for improvements in workplace conditions.[43] Settlement house residents also advocated for the creation of parks and playgrounds where working-class and immigrant families could escape the danger and filth of city streets, as well as child-labor laws and childcare facilities that reduced the temptation for desperately poor families to send their children to work in factories. Over time, settlement workers also became increasingly engaged with municipal politics, recognizing that little could be done to improve the urban environment as long as corrupt municipal officials remained in thrall to the owners of saloons, brothels, tenements, and factories who benefited from the exploitation of the poor and powerless.[44]

When urban environments were judged irredeemable, reformers sometimes sought to eliminate them entirely. In the early 1890s, for example, Hull House leased a parcel of land and demolished the dilapidated houses that had been built on it to make room for a park and playground.[45] In other cases, when it proved impossible to either ameliorate or eliminate a dangerous, unhealthy, or immoral urban environment—whether it be an unsanitary tenement, a hazardous factory, a rowdy saloon, or a seedy brothel—they instead sought to remove vulnerable individuals from that environment or to prevent them from encountering it in the first place. With an eye toward the protection of the health and reputations of young women from the surrounding districts, for example, Hull House residents—many of whom, including Addams, were unmarried or divorced themselves—offered counseling, training, and alternative forms of leisure. In these ways, social reform as practiced by the settlement movement navigated between a commitment to practical democracy and a desire to promote specific values that reformers associated with social progress. It entailed helping people to choose the environments that reformers believed were best for them while also reshaping the environments that were available for them to choose.

The settlement movement was just one of number of initiatives in the United States and in Europe that sought to solve the social problems of the industrial city by improving the conditions of everyday life.[46] The town planning movement, for example, embraced some of the same aims and methods as the settlement movement, although it operated on grander scale and placed greater confidence in experts. The Scottish biologist and social reformer Patrick Geddes, for example—one of the leading international advocates of town planning—had in 1884 helped establish an "Environment Society" that aimed to improve the conditions of the poor in Edinburgh's Old Town.[47] In 1899 and 1900, Geddes visited Hull House to learn from Addams and her fellow residents and to share his theories of how urban reform could promote the biological and social development of its inhabitants.[48] Closer to home, the Chicago-based town planner Daniel Burnham, architect of the 1893 World's Columbian Exposition and author of the influential 1909 *Plan of Chicago*, approached the city as "an organism in which all the functions are related one to another."[49] Only very partially implemented over the succeeding decades, Burnham's 1909 plan expanded to a grand scale some of the same social reform impulses that drove the settlement movement, though stripped of their most democratic and pragmatic elements.

By focusing on the ways in which the settlement movement adopted and adapted the concept of environment to its own practical efforts, we can see better both what it had in common with other progressive reform efforts and what made it distinctive. What it shared with town planning and social welfare policies was the conviction that the social problems of the late nineteenth and early twentieth centuries could be solved by improving the fit between urban residents and their industrial surroundings. What distinguished it was a focus on democratic, participatory, situated, and pragmatic reform. In itself, the idea that society could be best understood as an "organism" in relation to its "environment" could be used to support both top-down town planning exercises such as Burnham's *Plan of Chicago*, which sought impose a unified vision of social order on a chaotic city, and pragmatic experiments such as Hull House, which attempted to collaborate meaningfully with poor and working-class immigrants and to respond to their needs and wishes. The idea that society was an organism could, in other words, be deployed both by those who believed that they could mold that organism from the outside and by those who believed that progress would come only when the entire organism, from its lowliest to its most elevated members, worked together to improve its collective lot. In this way, the progressive discourse around the social environment was politically ambiguous, leaving open the question of how the social organism could be improved and by whom.

THE EVOLUTION OF SOCIETY

The settlement movement emerged at a time when practitioners in fields such as medicine, psychology, and sociology were beginning to explicitly reframe their goals and problems in terms of "environments" and the entities they surrounded. Many of them drew directly on the concept of environment as it had been articulated in the grand theories of Comte, Spencer, and their followers.

Addams read widely in the social theory of her day, and in a later account of her tour of Europe in the mid-1880s, she recalled spending one late night trying to formulate her dreams of meaningful social action in "ill-digested phrases from Comte."[50] While Comtean positivism failed to win Addams's allegiance, the faith in the intentional improvement of the human condition through the application of science that it represented influenced the path she ultimately took.[51] Even more important was the evolutionary theory of the day, which was included as part of the educational program of Hull House; some of the most popular lectures delivered in its early years, Addams later recalled, belonged to a series on the theme of "organic evolution" delivered by a young university lecturer.[52] Other settlement house residents picked up their understanding of evolutionary theory in less formal ways, but they shared Addams's faith that the settlement movement could contribute to an ever-closer, more refined, and more just coordination between the social organism and its conditions of existence.

The particular version of evolutionism embraced by the settlement movement built on but also challenged the liberal, laisser-faire variety advocated by Herbert Spencer, who had been largely responsible for introducing the term *environment* into English-language scientific discourse in the 1850s. Extraordinarily influential in the late nineteenth century, Spencer's vision of evolution as the progressive adaptation of organisms to their environments helped naturalize Victorian ideas of social and civilizational hierarchy across the English-speaking world. Beginning with his 1855 *The Principles of Psychology* and continuing in a series of works published over the succeeding decades, Spencer argued that each organism should be understood as constantly seeking to adapt to its "environment." Organisms and human societies alike, he argued, could be evaluated and ranked by the complexity and efficacy of those adaptations as they evolved through time. As he put it in *The Principles of Sociology*, although "the characters [i.e., characteristics] of the environment co-operate with the characters of human beings in determining social phenomena," the precise nature of the cooperation differed as one rose up the scale of social evolution, with more advanced civilizations showing both more sensitivity to and more control over their surroundings. Societies that were "highly organized, rich in appliances [i.e., technologies], advanced in knowledge, can, by the help of various artifices, thrive in unfavourable habitats, yet feeble, unorganized societies cannot do so: they are at the mercy of their surroundings."[53]

While the settlement movement generally accepted Spencer's idea of social evolution and the hierarchies it implied, it rejected the radical liberalism that led him to argue that any attempt to intentionally reshape society through government could only slow the process of adaptation. On the contrary, Addams and her colleagues believed that modern, industrialized societies had reached such a level of complexity and interdependence that further progress was possible only through intentional, coordinated planning. In support of this belief, they drew on the theories of post-Spencerian progressive evolutionists of the late nineteenth century such as the American botanist and sociologist Lester Ward, who distinguished between a "genetic" form of adaptation that operated unconsciously and automatically in all forms of life and a "telic" one through which humans consciously sought to achieve their goals.[54] For Ward, the fact that some of the plants introduced to North America had grown even more vigorously in their new homes than in their native habitats proved that evolution did not automatically produce a perfect fit between an organism and its environment. On the contrary, organisms had latent potential that was not expressed in their ordinary environments. Spencer's use of evolutionary theory to justify laisser-faire liberalism was therefore contradicted by the study of nature itself, which revealed that relationships between organisms and their environments could be dramatically and intentionally improved through human action. The implications for social policy were clear: rather than allowing evolutionary processes to operate unchecked, humans had a responsibility to "accelerate social evolution" for the sake of the "conscious improvement of society by society."[55]

Although neither Ward's overarching theoretical framework nor many of his specific arguments and examples were convincing to most US sociologists as the field professionalized at the end of the nineteenth century, his interest in the biological foundations of society and in the possibility, contra Spencer, of intentionally accelerating social evolution left a deep impression. For Small and other sociologists in Chicago, for example, the study of living beings and the study of society were part of a single structure of knowledge, the latter modeled on the former but not reducible to it.[56] What brought the two into alignment was the generalized concept of the organism, which could be applied to individuals as well as societies. As Small and Vincent explained in their 1894 textbook *An Introduction to the Study of Society*, the idea that "society is an organism" was fundamental to sociology, since it made it possible to identify the distinct functions played by each part of society in relation to the whole.[57] They hastened to point out that human societies were not literally organisms in the sense that animals or plants were; nor did human societies, considered as organisms, necessarily contain all the kinds of parts or display all the functions that biologists had identified in living beings.[58] Just like those living beings, however, society could also be considered a "whole whose parts are intrinsically related to it, which develops from within, and has reference to an end which is involved in its own nature."[59] The task of sociology, they believed, was to understand this organism so that it would be better able to improve itself.

The settlement movement created its own distinctive form of progressive evolutionism by combining the ideas of Comte, Spencer, Ward, Small, and others with the practical experience of social reform. Addams, for example, in addition to being inspired by her reading of Comte in the 1880s, was in close conversation throughout the 1890s and 1900s not only with Small but also with the pragmatist philosopher and progressive reformer John Dewey, who taught at the University of Chicago from 1894 to 1904 and regularly visited Hull House.[60] Like Dewey, she embraced aspects of evolutionary theory even while remaining critical of evolutionists who neglected the agency of the individual or presumed that morality was somehow in conflict with biology.[61] Organisms did not simply adapt to the threats and opportunities posed by their surroundings, she believed; they also transformed those surroundings in ways that determined their own future development. This was particularly the case for human beings, whose psychological and social life was so much richer than that of other species. In light of the "modern evolutionary conception of the slowly advancing race whose rights are not 'inalienable,' but hard-won in the tragic processes of experience," Addams wrote in her 1906 book *Newer Ideals of Peace*, it would not suffice for humanity to allow natural evolutionary processes to work unchecked, as Spencer and his disciples had argued.[62] On the contrary, echoing Ward, she proposed that progress would result from understanding and controlling the psychological and social processes that mediated the organism–environment relationship.

Progressive evolutionism also provided a framework for understanding and justifying the leading role that women played in progressive reform. In her 1898 book *Women and Economics*, Hull House resident Charlotte Perkins Gilman built on Ward's call for a progressive, post-Spencerian sociology that would accelerate human evolution by empowering women.[63] The moral and social problems of modern society, she argued, could be traced to an economic system that made women dependent on men rather than on their own labor—indeed, one that made men into the "immediate and all-important environment" of women, to which women were forced to adapt in order to survive.[64] This unequal "sexuo-economic relation," as Gilman called it, slowed not only the evolutionary progress of women as a sex but also the race as a whole: "As we learn to see how close is the connection of that which we call the soul with our external conditions, how the moral sense and the behavior of man are modified by the environment, we must of course look for marked results in psychic development arising from so important a condition as our sexuo-economic relation."[65] Changing that relation, she argued, would transform aspects of the male and female character that had been presumed to be natural but were in fact products of the social environment. In a gendered interpretation of progressive evolution that drew inspiration from Ward, Gilman argued that social evolution would depend on what she saw as the most enlightened segment of society—namely, educated, independent white women such as herself—seizing the reins of the evolutionary process and directing it toward higher ends.[66]

Although some advocates of progressive evolution, including Gilman, used it to support their belief that certain races were more advanced than others, others saw evolutionary theory as a resource in their struggles for racial equality. Du Bois, for example, justified his social survey of Philadelphia's Seventh Ward by stressing its potential to reveal how members of different human races adapted to the social and industrial conditions they faced. As he explained in an essay on the theoretical framework of his research, the United States provided "the most remarkable opportunity ever offered" to study the influence of environment on human races.[67] Unlike mid-nineteenth-century race theorists such as Robert Knox and James Hunt, however, Du Bois focused on the "social environment," particularly the pervasive racial discrimination that he believed had delayed and distorted African American adaptation to the conditions of the industrial city.[68] Without entirely dismissing the possibility of deep-seated racial differences, Du Bois shifted the focus to the social conditions that aided or hindered the flourishing of an individual or a race. Because "in the realm of social phenomena the law of survival is greatly modified by human choice, wish, whim and prejudice," he wrote, "one never knows when one sees a social outcast how far this failure to survive is due to the deficiencies of the individual, and how far to the accidents or injustice of his environment"—a universal fact that was "especially the case for the Negro" in the United States.[69]

In the context of both Gilman's and Du Bois's varieties of progressive evolutionism, the progress of a sex, a race, or humanity as a whole depended mainly on the advances made by its most talented members. Du Bois famously placed his faith in the "talented tenth" of the African American population, while Gilman expected the "thinking women of to-day" to take responsibility for social evolution.[70] Other progressive evolutionists, however, embraced more egalitarian visions. Addams, for example, resisted attempts to transform the settlement house into a site where elite experts imposed their ideas on the supposedly ignorant masses or studied them as if they were specimens in a museum or laboratory. Such efforts, she argued, ignored one of the necessary burdens of democracy: "that we are bound to move forward or retrograde together."[71] By reframing social problems in terms of the relationship between the social organism and its environment rather than in terms of the moral failings of individuals, Addams hoped to convince other elites that such problems were indeed social in nature and therefore demanded social solutions, while also avoiding casting blame on the poor and powerless. Moreover, she believed that knowledge had to be developed together with practice rather than first being developed and then applied. As she wrote in 1899, "The ideal and developed settlement would attempt to test the value of human knowledge by action, and realization, quite as the complete and ideal university would concern itself with the discovery of knowledge in all branches."[72] In contrast to academic scholarship that sought to describe life in abstract terms, in other words, the settlement movement in general and Hull House in particular would "express the meaning of life in terms of life itself."[73]

By examining the way progressive evolutionism as advocated by Ward and other late-nineteenth-century sociologists was adopted and adapted by the settlement movement, we can see how a particular variant of the concept of environment was transformed from an abstraction to a concrete means of organizing action, and in the process shifted its form. Addams and her colleagues did not simply apply the concept of environment to their problems; rather, like the pragmatists they were, they put it into practice under particular circumstances and in the service of particular aims. Through the technique of the social survey and the institution of the settlement house, the idea of the "social organism" and its "social environment" took on a specific meaning, which was necessarily different from the set of meanings articulated by theorists such as Ward under other circumstances and with other aims. Both were environmental, but each was environmental in its own way.

FROM PROGRESSIVE EVOLUTION TO ECOLOGICAL ANALOGY

Although the settlement movement thrived from the 1890s to the 1910s, it lost momentum in the following years, as did the broader progressive movement of which it was a part. The settlement houses were partly victims of their own success, as some of their activities—from the construction and maintenance of small parks and playgrounds to the conduct of social surveys—were taken over by municipal and state governments or by academic sociology departments once they had demonstrated their promise. The playground that Hull House had built in the early 1890s after tearing down a row of dilapidated houses, for example, was turned over to the City Playground Commission after ten years.[74] Moreover, as the political winds shifted, the social ideals of the settlement movement were increasingly met with skepticism or even outright opposition. Between the United States' entrance into World War I in 1917 and the Immigration Act of 1924, which prohibited the naturalization of people of Asian descent and established strict quotas for others, Americans grew less sympathetic to the idea of devoting resources and expertise to improving the conditions faced by poor and working-class immigrants to industrial cities.[75] At the same time, the settlement movement began to fracture from within as a new generation of leaders with divergent visions came to the fore.[76] As a result, the movement's most influential and innovative period came to an end, with the number of settlement houses dropping to just over half what it had been in 1910 by the 1930s.[77] Hull House lasted longer than most, but in the wake of the Great Depression and Addams's death in 1935, it entered a period of turmoil from which it never fully recovered.[78]

The settlement movement also relinquished its place at the forefront of research on social processes and social problems as the discipline of sociology became increasingly professionalized and concentrated in academic departments. In the 1890s, the University of Chicago's sociology department had been closely engaged with the practical research and reform carried out by the settlement houses and with the results of the social surveys they performed. In the following decades, however, a younger generation of sociologists increasingly distanced themselves from practical reform and instead sought to establish identities as disinterested scientists. After Robert Park began teaching sociology at the University of Chicago in 1914, for example, his distaste for "do-gooders" became one of the department's defining characteristics.[79] Despite (or perhaps because of) his prior experiences as a journalist and as an assistant to Booker T. Washington at the Tuskegee Institute, Park was convinced that most efforts to directly reshape the urban environment to achieve social aims were doomed to failure.

Park therefore taught his students that sociology could best serve society by producing objective insights into the functioning of human groups, which would ultimately provide the basis for reforms that were more successful than those offered by the settlement movement. He and his colleagues in what became known as the Chicago School of Sociology consequently abandoned the pragmatic intertwining of knowledge and action that Addams and others had advocated. Instead, they endorsed a division of labor between sociologists concerned with the production of universal truths about human society, on one hand, and social workers and reformers concerned with the solution of immediate social problems, on the other. In 1920, this division of labor was formalized at the University of Chicago when it incorporated the previously independent Chicago School of Civics and Philanthropy as the central component of its new School of Social Service Administration.[80] Thenceforth, it was clear that reformist social work belonged in one part of the institution, scientific sociology in another.

In the 1910s and '20s, sociologists at the University of Chicago sought to establish a set of objective methods and professional standards that would keep them above the fray of partisan politics and the messy details of practical reform. Correspondingly, they argued that the social environment could and should be studied on its own terms, with techniques that were specifically tailored to the study of the social domain. While they continued to use the methods of the social survey, they focused most of their efforts on techniques they believed would give them a richer sense of the inner lives and social worlds of the city's inhabitants—that is, techniques that would produce data on aspects of human life in the industrial city that were purely social. For these purposes, what Park dismissed as the "trivial schedules" of the social survey and the inspections of living and working conditions carried out by settlement workers were of limited use.[81] Instead, he and colleagues such as William I. Thomas argued, sociologists would need to use letters and other personal documents, in-depth "life history" interviews, and what would later come to be known as participant observation—that is, sharing as much as possible in a community's existing patterns of work and life in order to understand them from the inside. An early example is the work of Nels Anderson, one of the graduate students under Park's tutelage, who spent a year living among Chicago's hobos while also collecting more than sixty life histories and "a mass of documents and other materials."[82] This expanded array of methods and sources aimed to document not the kinds of objectively determinable facts of the urban environment produced by the social survey—the number of saloons, the filthiness of the alleys, the crowding of tenements, the occupations and incomes of residents, and so forth—but rather what Thomas called the subjective "definition of the situation." It was that subjective definition, Thomas argued, that gave the observable facts their social significance.[83]

As they adopted these new methods, Chicago sociologists came to see the city primarily as a field site or even a "laboratory" for research—a metaphor that Addams had previously rejected as being insufficiently "human and spontaneous" to capture the collaborative, reform-oriented work of the settlement movement.[84] Within that laboratory sociologists focused their research on certain groups or urban types, such as the hobo, the gang member, the "unadjusted girl," or the "jack-roller" who made his living by robbing the indigent and intoxicated.[85] By reconstructing the life history of an individual belonging to one of these types, they aimed to reveal "the intimate interplay between his impulses and the effective stimuli of the environment."[86] Doing so often required researchers to become closely involved with the communities they were studying. Anderson, for example, had spent years riding the rails before he was recruited to study Chicago's hobos; he was in some ways more personally familiar with the people he was studying than any settlement worker had ever been, and he shared with the settlement residents the desire to conduct research that would aid in the project of social reform.[87] For Park, however, who supervised Anderson's research and wrote the preface to the 1923 book that resulted, *The Hobo: The Sociology of the Homeless Man*, the project's importance had nothing to do with reform. Rather, it was important because it showed how a detailed study of a particular community or subculture could produce insights into the fundamental processes that governed urban society. Such research was for Park primarily a contribution to "our permanent scientific knowledge of the city as a communal type"; only secondarily, if at all, was it aimed at improving the lives of the city's residents.[88]

As they consolidated their position as experts on "society," sociologists also gradually changed their position on the relationship between social and biological evolution. For late-nineteenth-century progressive evolutionists influenced by Comte and Spencer, the processes that governed human societies were intimately connected to the processes described by biologists, even if the two were nonetheless distinct. Society, they argued, was a refinement and a continuation of biology, not a domain entirely apart, and human society represented the highest and most complex result of evolutionary processes that were ultimately cosmic in nature. Sociologists of the Chicago School, by contrast, were dubious of the progressive evolutionary framework for multiple reasons. One was its grand scope, which was ill suited to the development of highly specialized disciplines, each focused on its own particular corner of the cosmos. Another was the way that nineteenth-century forms of progressive evolutionism had often been used to support racial and civilizational hierarchies, such as Spencer's attempt not only to distinguish between "highly organized" societies and "feeble, unorganized" ones but to depict such a distinction as natural. Against this tradition of evolutionary thought, Chicago School sociologists argued that society was an entirely autonomous domain governed by laws analogous to but fundamentally separate from those that biologists had identified for living organisms.[89] The idea of the "social organism" did not disappear from their work entirely, but it became increasingly metaphorical, losing even the tenuous links that it once had to the study of biological organisms in the early days of Chicago sociology.

Thus, while Chicago School sociologists borrowed the language of plant ecology to describe the "processes of competition, invasion, succession, and segregation" affecting urban communities, they saw no need to study the city's soils or vegetation, or even the kinds of material factors such as crowding and sanitation that Addams and her colleagues had tried to track through social surveys.[90] Rather, what mattered most to Park and his colleagues was the "social environment" or the "moral climate," which was defined by urban residents' subjective perceptions of the social forces and actors surrounding them—that is, Thomas's "definition of the situation."[91] Their aim in borrowing the language of plant ecology was to gain access to a rich set of metaphors and to bolster their scientific authority, not to pursue studies that linked human society to its material surroundings.[92] The increasingly sharp divide between biology and sociology also had implications for their view of social reform. If what mattered were not the material conditions of the industrial city but rather residents' perceptions, attitudes, and values with regard to those conditions, then social problems could not be solved simply by closing saloons, building playgrounds, cleaning streets, or renovating tenements. On the contrary, Park wrote in a 1925 essay on juvenile delinquency, it was clear that "any effort to re-educate and reform the delinquent individual will consist very largely in finding for him an environment, a group in which he can live, and live not merely in the physical or biological sense of the word, but live in the social and psychological sense."[93] It was therefore the "play group" as a social unit that ultimately mattered, not the playground as a material environment.[94]

Understanding society and its surroundings in this way helped shift the focus of research and reform from the everyday municipal and domestic concerns of the settlement movement to the exchange of ideas and information in complex modern societies. For Park and his colleagues in the Chicago School, it was clear that modern societies were structured through formal regulation, large institutions, and the mass media of newspapers, films, and radio rather than through the kinds of personal relationships and face-to-face encounters that the settlement movement had cultivated.[95] In a mass society of this kind, they argued, the most pressing threats were not unsanitary alleyways or dangerous factories but the breakdown of the social fabric—something that could be accelerated by the efforts of businesses and governments to reshape the social environments of urban residents in ways that heightened profits and concentrated power but undermined democracy and social order. Changing perceptions, attitudes, and values therefore took higher priority than changing the material environment, and new institutions and methods were required to do so. These included what Park called "social advertising," which sought to "educate the public and enlist the masses of the people in the movement for the improvement of conditions of community life" by exposing them to certain images, ideas, and narratives.[96] In this new world, the "city environment" or "urban environment" was still important, but it was an environment that consisted of subjective perceptions and values rather than of pragmatic and embodied engagements with one's social and material surroundings.[97]

In the years following World War I, critics of the progressive movement used this new way of defining and materializing the environment to question the very possibility of democratic, pragmatic social reform as it had been advocated by Addams, Dewey, and other progressives. The journalist Walter Lippmann, for example, based his widely debated critique of US democracy on the existence of an unbridgeable gap between the subjective world of attitudes and beliefs of individuals and the objective world of social and physical realities—a position that led him to propose delegating some of the most important social decisions to a technocratic elite rather than to the kind of pragmatic experimentation pursued by the settlement movement. In his 1922 book *Public Opinion*, he argued that individuals in modern mass societies lived in "pseudo-environments" consisting of understandings of the world around them that were at best partial and at worst highly inaccurate.[98] These pseudo-environments consisted of nothing but "pictures inside our heads," which became "Public Opinion" when they were transformed into the basis of collective action.[99] In complex modern societies, however, Lippman argued that it was impossible for any individual to have a firm grasp on all matters of public concern, which meant that such "pictures" would always be faulty. Reformers who placed their faith in the participation of an informed public in democratic policymaking were therefore bound to be disappointed.[100] Whereas Addams and other pragmatic, progressive evolutionists had argued that social progress depended on creating institutions where people of diverse ethnicities and classes came together to pursue both knowledge and action, Lippmann suggested that only experts and the public administrators who employed them could be trusted to gain an understanding of the "real environment."[101] In an era of mass media and disciplinary specialization, such arguments became increasingly compelling to many Americans, overshadowing the previous generation's faith in the possibility of improving the social organism by transforming its environment.

By following the rise and fall of a variety of progressive politics that sought to improve the health of the social organism by reforming its environment, we can see how the ability of a concept to make a difference in the world depends on more than the content of the concept alone. The power of the variant of the concept of environment that had been articulated by progressive evolutionists and settlement reformers did not vanish overnight; indeed, there are aspects of it that remain appealing today. What changed was the world in which that concept could be put into practice, including the specific techniques and institutions that made it visible and tangible. As sociologists professionalized at the University of Chicago and elsewhere in the early decades of the twentieth century, for example, the allure of a concept that linked them tightly to non-professionals and nonacademic contexts such as the settlement house faded. Similarly, as the social structure of the industrial city shifted in response to new forms of mass media and mass consumption, the neighborhood-based environmental reforms advocated by the settlement movement came to seem—and in some cases actually to be—less effective than efforts to change urban residents' "definition of the situation" through "social advertising" and other means. As the world changed, this variant of the concept of environment, which had been both inspiring and effective for decades, suddenly lost its appeal.

*

In the summer of 1892, when progressive evolutionism was at its height and Hull House was leading the US settlement movement into its most innovative and expansive period, Addams was invited to lecture at the School for Applied Ethics in Plymouth, Massachusetts. In the published version of her lecture, which summarized the aims and objectives of Hull House as they had developed over the previous three years, she warned that the "one thing to be dreaded in the Settlement is that it lose its flexibility, its power of quick adaptation, its readiness to change its methods as its environment may demand."[102] In other words, she suggested, the settlement house was a kind of organism that would thrive not simply by responding to the demands of its environment but also by maintaining its capacity to respond to the still-unknown environments of the future. In this way, Addams wrote, the settlement house was an "experimental effort to aid in the solution of the social and industrial problems which are engendered by the modern conditions of life in a great city."[103]

The experimentation embodied in Hull House and other settlement houses of the period was linked to a view of society as an organism that evolved over time in relation to an environment that consisted of both physical and social factors. This evolution was not, in Addams's view, something that happened by itself, as Spencer and his disciples had argued. Nor was it a process that would inevitably reach some foreordained conclusion. On the contrary, social evolution had to be carefully nurtured and directed by humanity with the interests of the social organism as a whole in mind. The only question was how the process of evolution was to be directed. For Addams, the answer to that question was the settlement house, an experiment that brought middle-class reformers and working-class immigrants together without a predetermined outcome but instead with the aim of discovering how best to improve the lot of all. The social survey was an important tool in this process of discovery, bringing to light regularities in the relationship between social disorder and the conditions of the industrial city that could serve as the basis of reform. Just as important, however, was the day-to-day process of collaboration and conversation, which Addams believed produced both stronger communities and a more robust knowledge of society.

Over the course of the 1910s and 1920s, both the progressive evolutionism and the pragmatic social reform advocated by Addams were largely abandoned by US sociologists, particularly those of the Chicago School, who used stark divisions between society and biology to simultaneously distance themselves from the racist implications of progressive evolutionary theories and bolster their authority as scientists. Ironically, the use of ecological analogies by Park and others reinforced this division, even as it strengthened their commitment to an "environmental" account of society. In their work, the urban "environment" to which the "social organism" responded had little to with the physical structure of the city or the living beings found within it. Rather, it was purely social and psychological in character—that is, connected to the material environment only through the perceptions, attitudes, and values that were subjectively attributed to that environment by residents of the city as part of their "definition of the situation." Consequently, Park and his colleagues argued, social problems arose from social disorganization and disorders of personality rather than from crowded tenements, unsanitary alleyways, raucous saloons, or other aspects of the material environment of poor and working-class Chicagoans. To study such problems, they therefore largely replaced the social survey with a new set of methods capable of revealing the inner experiences and social relations of the urban resident.

By examining the environmentalism of the settlement movement, we can see how the adoption and adaptation of explicitly environmental language was closely linked to a particular social and material context—in this case, the industrial city and its problems. It was in the industrial city, reformers and researchers such as Addams, Du Bois, and Park believed, that human beings faced an unprecedented environment fashioned almost entirely by human hands. Combined with an influx of immigrants from Europe and elsewhere and internal migrants from rural areas within the United States, this novel environment presented a unique opportunity to study and perhaps to improve the relationship between the structure and function of the social organism and its conditions of existence. Like naturalists in Paris at the end of the eighteenth century or physicians in the British Empire in the mid-nineteenth century, urban reformers and sociologists in the United States at the beginning of the twentieth century crafted an environmentalism suited to their aims and circumstances. In doing so, they sought to create new forms of solidarity among the diverse peoples living under urban industrial condition, and in a few cases actually succeeded. The next chapter explores another set of environmentalists who similarly grappled with the consequences of industrialization, but did so on the scale of entire nations and even the planet itself.

CHAPTER 4

The Biosphere as Battlefield: Strategic Materials and Systems Theories in a World at War

Tungsten is a hard, dense metal with an extraordinarily high melting point—distinctive properties that by the beginning of the twentieth century had brought it into demand for the manufacture of high-speed machine tools, including those used to produce arms and ammunition. However, the very same properties also made it so technically challenging both to purify and to alloy with other metals that the international market for processed tungsten and tungsten alloys was dominated in the early decades of the twentieth century by a single industrially and scientifically advanced nation, Germany. The consequences of this dependence became apparent with the outbreak of World War I in the summer of 1914, when Germany's enemies suddenly found themselves cut off from a critical strategic resource.[1] While the United Kingdom and the United States were able to expand their own production of refined tungsten, other nations lacked the necessary expertise and industrial facilities to do so. Russia, for example, despite having abundant deposits of tungsten ore within its borders, entirely lacked the capacity to refine the metal, forcing its manufacturers to seek out foreign sources. By the end of the war, they were importing about 3,500 tons of ferrotungsten alloy from Britain annually.[2] With prices rising to stratospheric heights even as global production doubled to meet the demands of the war, this was a strategy that proved both expensive and unreliable.[3]

Tungsten was only one of many materials that became scarce or expensive in Russia as a result of the war, with resulting disruptions to its economy and to its ability to defend itself against its enemies. In 1915, the Russian Academy of Sciences responded to this crisis by establishing a Commission for the

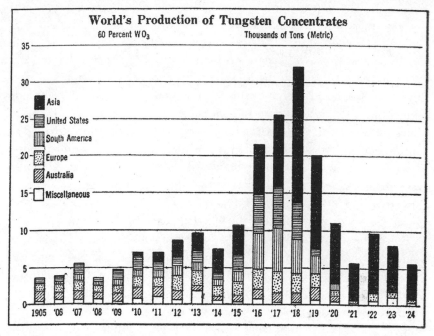

FIGURE 8. A chart showing the increase in the world's production of refined tungsten concentrates during World War I (1914–1919). (Reprinted from figure 4 in Josiah Edward Spurr, "Steel-Making Minerals," *Foreign Affairs* 4, no. 4 [1926]: 601–12, on 610, with permission from the Council on Foreign Affairs, conveyed through Copyright Clearance Center.)

Study of the Natural Productive Forces of Russia, which aimed to identify domestic sources of strategic materials and to develop the methods and expertise to exploit them.[4] Usually known by its Russian acronym KEPS, the commission was the brainchild of the mineralogist Vladimir Ivanovich Vernadskii, who saw his country's inability to exploit its own resources as a symptom of a deeper malady that would continue to afflict it even after the war and that could be cured only through the development of Russian science.[5] As he wrote in 1915, "One of the consequences—and also one of the causes—of Russia's economic dependence on Germany is the extraordinary insufficiency of our knowledge about the natural productive forces with which Nature and History has granted Russia."[6] KEPS sought to make up for that insufficiency through research, including numerous expeditions in search of deposits of tungsten, tin, aluminum, and other minerals essential to the war effort.[7]

While Russia was more dependent on foreign imports than many of the other nations involved in World War I—a dependency that was particularly striking in light of its vast territory and abundant natural resources—it was not alone in facing critical shortages of what were then coming to be known as "strategic materials."[8] In many other nations, politicians, industrialists, military leaders, scientists, and engineers helped create expert commissions like KEPS in the hope of reducing their own dependence on foreign imports.[9] Diverse in scope and institutional form, these commissions had a common concern with the material basis of national survival under the conditions of global warfare. For the scientists who participated in them, they provided an unprecedented opportunity to survey the resources of their nations and of the globe as a whole in relation to industrial needs, as well as to develop new ways of organizing manufacturing and trade. Such wartime work was a transformative experience for many of these scientists, setting their research on a new trajectory over the course of the following decades.

Few of the scientists involved spoke of their nations' economies and the resources on which they depended in terms of "organisms" or "environments." Nonetheless, the data they gathered and the models they developed during World War I provided a foundation for new ways of understanding the concept of environment in the postwar years. In particular, scientists involved with or influenced by these wartime commissions increasingly adopted mathematical and quantitative techniques for modeling "systems"—that is, collections of living and nonliving components bound together through flows of energy and materials. Like organisms, they argued, systems were organized assemblages of diverse parts that could only be understood in relation to the conditions that surrounded them. The value of Russia's tungsten ore deposits, for example, could not be assessed in isolation; it depended both on the nation's industrial capacity to extract and purify them and on the price and availability of tungsten on the international market. More broadly, rather than seeing resources in terms of absolute quantities, fixed uses, and intrinsic values, scientists came to see them as sources of matter or energy whose value, significance, and abundance were determined by the demands and capacities of industrial systems and by the vagaries of geopolitics and international trade.

Wartime efforts to inventory such industrially important resources and to maximize the efficiency and productivity of national economies also shaped the ways scientists understood the environments of living beings. In the decades following World War I, ecologists, demographers, and geochemists applied techniques and concepts developed to manage strategic materials during the war to the study of exchanges of matter and energy between organisms and their environments. In the 1920s, for example, Vernadskii and his students developed new methods for measuring biogeochemical processes on local, regional, and global scales. These processes, Vernadskii argued, collectively constituted Earth's "biosphere," whose structure and function depended upon its relation to the planet's "cosmic milieu." In the following years, a number of other scientists, including the US ecologist G. Evelyn Hutchinson and his students, extended these ideas and methods to study flows of energy and matter that linked living beings and nonliving matter into what were then beginning to be called "ecosystems." After World War II, ecologists working in the shadow of the atomic bomb adopted similar concepts and methods to reveal how human activities were transforming the planet. In these ways, the ecological and conceptual legacy of two global wars stretched into the second half of the twentieth century, reshaping how scientists understood the relationship between life and its surroundings.

MOBILIZING

In the half-century before World War I, the world's economies had been linked together through an expanding global infrastructure of transportation, communications, and finance, which made it possible for manufacturers in one country to take for granted the regular delivery of essential materials and products from other countries. This thickening web of trade was ripped apart by the outbreak of World War I, which erected new barriers ranging from legal embargoes to naval blockades. At the same time, the conduct of the war was heightening demand for a wide range of industrially produced goods, from food to bullets. Particularly on the western front, where the war was fought and won largely through attrition rather than through tactical or strategic brilliance, the main challenge for the combatant nations was producing and delivering men and materiel to the front lines at rates sufficient to make up for unavoidable losses.[10] As became apparent soon after the war began, the nation most likely to emerge victorious under these conditions was not the one with the most advanced weaponry or the most skilled generals, but the one capable of maintaining and increasing the production of industrial goods despite wartime disruptions and demands.

Perhaps the most basic of these industrial goods was food, the production and distribution of which became a central concern of all the combatant nations. Over the course of the war, each nation faced the challenge of delivering food to thousands or even millions of soldiers while continuing to provide for civilians at home. The challenge was heightened by the fact that soldiers on active duty on the front lines generally consumed more food than they had as civilians, even as the ability to produce that food was being compromised by the conscription of large numbers of farmers, the transformation of productive farmland into battlefields, and the disruption of the usual systems of food distribution. Making matters worse, droughts, floods, and other extreme weather events in 1916 and 1917 led to lower-than-usual harvests in many parts of the world.[11] Even in the United States—a late entrant to a war fought on other nations' territories—a food crisis emerged as it became clear that its European allies "must have more food than we can raise, and we must send them more than we can readily spare," in the words of a government pamphlet published in 1917.[12] Accordingly, the US federal government implemented new measures to expand the amount of land being farmed, encourage the adoption of mechanized farming equipment, increase the use of chemical fertilizers, facilitate food conservation, and promote community and home gardening.[13] By 1919, these measures had resulted in, among other things, an estimated 240 million acres of cereal grains being planted—an increase of 33 million acres over the prewar period that yielded 625 million additional bushels of grain.[14] Even so, food shortages continued to afflict combatant nations.

In agriculture as in other domains, the war did more than disrupt trade. Through the intentional and unintentional destruction of farms, forests, mines, and factories, it also directly obstructed the extraction and processing of resources. In some cases, apparent declines in production were illusory, reflecting merely the fact that one nation's forces had occupied sites of extraction, processing, and manufacturing belonging to an enemy and redirected their products to their own war needs. In other cases, however, actual rates of extraction or production slowed or stopped entirely, whether because the necessary labor could not be mustered, ongoing combat made operations impossible, or occupying forces had stripped mines and factories of valuable equipment or destroyed them before they could be reclaimed by the enemy. In northeastern France, for example, the production of coal, steel, lead ore, and iron ore dropped dramatically during the war, even allowing for the appropriation of French mines and factories by German forces. As early as late 1914, the Germans had begun removing steam engines, electric dynamos, railcars, locomotives, stamping presses, and other mining equipment from the border departments of Nord and Pas-de-Calais.[15] Later, as the end of the war approached and defeat seemed inevitable, they also began destroying mines, factories, and infrastructure to prevent them from falling into the hands of the Allies. One observer estimated that a total of 124 coal mines, 500 miles of mine tracks, and more than 16,000 miners' houses had been destroyed in the Nord and Pas-de-Calais coal fields alone.[16] Meanwhile, trench warfare was transforming vast swaths of northeastern France into wasteland, with entire forests "blown into splinters by shellfire."[17]

Even when the extraction and processing of natural resources was not directly disrupted by combat, trade embargoes and naval blockades could lead to critical shortages, as was the case with timber in the United Kingdom. Like food, wood was quickly recognized to be one of the war's most important strategic materials. At home it was essential for propping up the tunnels of coal mines, making crates for munitions, and innumerable other civilian and military purposes, while on the front it was used to build barracks, trenches, fencing, railroad tracks, and telegraph poles.[18] In principle the United Kingdom had access to sufficient timber for wartime needs throughout its vast empire and through trade with Russia, Sweden, Norway, the United States, and Canada, all of which had abundant forests and productive timber industries located at safe distances from the war's main battlefields. In practice, however, Britain faced a severe timber shortage in 1916 as a result of German submarine warfare, limited shipping capacity, and rising demand. In response, it began harvesting its domestic forests at an unprecedented rate.[19] As early as 1916, the British forester Edward Percy Stebbing noted that the country had already begun "sacrificing considerable areas of young woods and felling old ones of any value, since we must supply the urgent needs of the country."[20] By April 1917, about 100,000 acres in the British Isles had been clear-cut, and the rate of cutting continued to accelerate over the remainder of the war.[21] According to one estimate, about 450,000 acres, or half of the country's productive forested land, had been laid bare by the time peace was declared in November 1918.[22]

When foreign sources of a particular strategic material were unavailable and there were no domestic sources waiting in reserve to be exploited, nations with the necessary industrial capacity and expertise sometimes turned to natural or synthetic substitutes. In Germany, for example, scientists and industrialists had been warning of the risks of dependence on imported nitrates even before the outbreak of World War I, but the war provided the motivation and resources to develop a domestic substitute on an industrial scale. After the country lost access to Chilean nitrates that it used to manufacture both fertilizers and explosives, it began building factories to "fix" atmospheric nitrogen using a process developed by the chemists Fritz Haber and Carl Bosch.[23] In 1915, a pilot plant near Ludwigshafen began producing nitrates for the manufacture of explosives; by the spring of 1917, a second plant capable of fixing 130,000 tons of nitrogen per year was operating near Halle.[24] By the end of the war, the amount of nitrogen fixed using the Haber–Bosch process was equal to the total amount used in Germany before the war.[25] Fixing nitrogen required an enormous amount of energy, but with some of Europe's richest coal mines lying within Germany's borders, that was a resource not in short supply.[26]

Once the war was over, some of these new forms of resource extraction and domestic production expanded even further. The amount of nitrogen fixed using the Haber–Bosch process, for example, climbed in the decades after the war, transforming not only the international trade in nitrates but also the practice of agriculture and the operation of the global nitrogen cycle.[27] In other cases, wartime booms proved ephemeral. Many of the US farmers who had expanded onto marginal lands and taken out loans to invest in farming machinery went bankrupt in the early 1920s when the high prices and easy credit they had enjoyed during the war years evaporated.[28] In the United Kingdom, meanwhile, the restoration of the international timber trade provided a respite for domestic forests. In any case, whether wartime initiatives faltered or flourished with the return of peace, they had lasting effects. British foresters, for example, launched an aggressive domestic program of afforestation to supply the nation's future needs while also seeking to expand and rationalize the exploitation of timber resources throughout the empire.[29]

By approaching World War I in terms of its effects on the production and circulation of strategic resources rather than in terms of military strategy or the soldier's experience of warfare—that is, by focusing on what the director of the US Geological Survey described at the war's end as the "strategy of minerals"—we can see how the war reshaped the world that scientists sought to describe, provided new opportunities and motivations for them to study that world, and focused their attention on certain of its aspects rather than others.[30] In particular, the enormous demands of the war for raw materials and finished goods, coupled with its disruptions to systems of production and trade, called scientists' attention to the need for techniques to quantitatively assess the amount of resources available and to track their movements and transformations from one part of the economic system to another. We can see, in other words, how the aim of sustaining economies that were globalized, industrialized, and at war with each other drove scientists to embrace a very specific set of tools for understanding the relationship between a nation's economic and military survival and the external conditions it faced—tools that would, in the postwar period, be applied to a much broader range of questions.

TAKING STOCK

The establishment of wartime resource commissions such as KEPS was driven by the conviction that industrial production under the conditions of modern warfare required centralized government coordination and, moreover, that such coordination, dependent as it was on a detailed understanding of complex scientific and technical processes, could not succeed without the advice of experts. In practice, these commissions were often underfunded and largely ignored by politicians and industrialists. They nonetheless proved transformative for the scientists who participated in them. Motivated by their new roles as wartime advisors to governments, these scientists pursued new kinds of research and developed new kinds of models to understand "resources" as diverse as mineral deposits, arable farmland, and human populations. Under the urgent conditions of war, scientists, engineers, and economists not only sought to measure various resources and to map their distributions across Earth's surface, but also to determine their accessibility and value in relation to changing levels of supply and demand, the availability of substitutes, new methods of extraction and processing, and economic and geopolitical constraints.[31] After the war's end, they continued to develop techniques and theories they had embraced during the war and to expand their scope of application. Central to this war-inspired research was an approach to evaluating resources that assumed the value and uses of any given resource could be assessed only in relation to the total "system" of which it was a part, as well to as the environment in which that system operated.

Concerns about resources and their availability predated World War I; after all, resource shortages could and did arise for many reasons besides war, and the process of industrialization and the expansion of international trade over the course of the previous century had forced scientists, engineers, and experts in political economy to grapple with the possibility and consequences of such shortages. As early as the 1860s, the English political economist William Stanley Jevons had called attention to the intricate webs of international trade that kept the British industrial economy humming, as well as its profound dependence on one particular resource, coal.[32] Nonetheless, the war intensified and expanded such concerns, in many cases transforming speculative possibilities into harsh realities. In the case of nitrogen, for example, growing concern over Germany's dependence on imports during the decade or so preceding the war had led scientists to develop experimental techniques for producing ammonia and nitrate from atmospheric nitrogen. It was only after the outbreak of war, however, that German politicians agreed to invest the enormous resources needed to make it a practical alternative to Chilean sources.

The shift in the assessment of resources toward an increasingly systems-oriented view can be seen in the work of KEPS, which sought to determine whether or not Russia possessed deposits of industrially important minerals as well as what forms of expertise and industrial capacity would be needed to make those deposits useful. In the case of aluminum, for example, Russia did not face quite the same crisis as it did in regard to tungsten, since the international market for the former was dominated by two of its allies, France and the United States. Nonetheless, aluminum prices rose dramatically as a result of the war, more than tripling by 1916.[33] Recognizing the threat that Russia's dependence on imported aluminum posed to its industrial capacity and military strength, KEPS made the search for exploitable deposits of aluminum-bearing ore one of its early areas of emphasis. In the spring of 1915, Vernadskii and Alexander Fersman, a former student of Vernadskii's who served as secretary of KEPS, wrote to the Ministry of Trade and Industry to propose establishing a new aluminum industry on the basis of recently discovered Russian bauxite deposits.[34] Although the proposal stagnated during the war, it sparked an interest in domestic sources of aluminum that eventually led Fersman to identify deposits in Russia's far northwest in the 1920s that became the basis of an important industry.[35] More broadly, the pressures of war helped convinced not only the czarist government that fell in 1917 but also the Bolshevik government that succeeded it that the advice of scientists was essential to economic and military survival.[36]

Wartime shortages led even officially neutral nations such as Norway to scour their landscapes for materials that, with the help of new industrial processes, would allow critical industries to continue. Resources that might have seemed insufficiently valuable to exploit under ordinary conditions—such as difficult-to-access mineral deposits, low-quality stands of timber, and marginally productive cropland—became the targets of intensive investigation. In Norway these efforts were led by the mineralogist Victor Moritz Goldschmidt, who was appointed chairman of the State Raw Materials Commission and director of the Raw Materials Laboratory in 1917. As Goldschmidt wrote in a survey of Norwegian and foreign research in 1918, the outbreak of war had made the importance of "industrial independence" obvious. Such independence, he argued, could be established only through research on as-yet unutilized minerals as well as the new refining and manufacturing processes that would allow them to substitute for foreign imports.[37] The aim of the Raw Materials Commission and Laboratory was thus to identify domestic sources of aluminum, phosphorus, potassium, titanium, and other elements essential to Norwegian manufacturing and agriculture.[38] Such work continued into the postwar years, when Goldschmidt identified olivine—a magnesium-rich mineral found in abundance on Norway's western coast—as a useful refractory material for foundries. Eventually, the work of the Raw Materials Laboratory helped Norway establish a world-leading olivine industry on the basis of this once-neglected mineral.[39]

Simply identifying and characterizing such resources was not enough to prove their value, as Goldschmidt had discovered in his initial, failed attempt to establish olivine as an economically viable domestic source of magnesium.[40] It was also important to assess them in relation to other potential resources and the broader economic system. For this purpose, scientists relied increasingly on quantitative measures of efficiency and productivity. In the United Kingdom, for example, domestic forests had long been assessed and managed to achieve a mix of aesthetic, recreational, and economic goals, including the maintenance of aristocratic hunting grounds, the production of pleasing landscapes, the protection of watersheds, and the harvesting of wood and other forest products for household or community needs. The decision to make up for disruptions to the international timber trade by harvesting domestic forests tipped the balance decisively toward quantitative measures of timber yield.[41] With roots stretching back to eighteenth-century Germany, the quantitative assessment of timber resources for the purposes of maximizing yield over the long term was not, in itself, novel.[42] What was novel was the deployment of such techniques to manage resources in relation to industrial needs on a national or even global scale. The role of the war in this shift was often made quite explicit. In 1916, Stebbing, the British forester, argued that the war had made it "a duty—a national duty—to see that every acre of land in this country is made to bring in the best return possible in the interests of the community as a whole."[43]

Partly because they were such powerful tools of abstraction, the reach of quantitative methods for assessing the value of resources in relation to available technologies and markets proved to be virtually unlimited. Even human populations could be treated as raw materials whose value depended on their relation to the needs and capacities of a given industrial system. Because the value of a human worker in such a system depended on specific skills and characteristics, simply counting the number of men or women of working age in a given population was not enough. Instead, experts devised increasingly refined techniques for characterizing the capacities of individuals and segments of the population and for determining the most effective strategies for what began to be described during the war as the "mobilization of human resources."[44] Rather than establishing a universal standard of quality, experts sought to determine precisely which members of the population were best suited to each of the specialized tasks demanded by industrial production and industrialized warfare, which were constantly changing as new technologies and tactics were devised. In the United States, for example, the army administered intelligence tests to more than 1.75 million recruits during the war to determine their fitness for particular ranks and duties, and it continued to assess them after recruitment through the Committee on Classification of Personnel.[45] Once soldiers had returned from the front, they continued to be treated as "human resources" that could be depleted through neglect or conserved through vocational training and rehabilitation programs.[46]

In the years following World War I, the apparent success of the centralized coordination of industrial production under the guidance of scientific and technical experts inspired technocratic movements on both the left and the right. To adherents of these movements, the complexity of modern societies and their dependence on international flows of materials and industrial goods meant that national survival depended on granting power directly to engineers, scientists, and other experts—or, if that proved politically impossible, on integrating scientific and technical expertise into the inner workings of the state. The form taken by these technocratic movements varied widely, however, in relation to each nation's domestic political systems. In the Soviet Union, the Communist government promoted technological megaprojects as a means to the rational exploitation of nature in the service of the proletariat (a project into which KEPS was integrated over the course of the 1920s).[47] In Weimar Germany and in the early years of the Nazi regime, meanwhile, revolutionary conservatives argued that the nation's defeat had resulted from its failure to mount a "total mobilization" of its industrial and human resources, which they intended to avoid repeating in any "total war" to come.[48] Technocratic planning of the use of resources also featured prominently in interwar visions of world government, in the developmental policies of European colonial administrations in Africa, and in the rise of the regional planning movement in the United States.[49]

By examining the development of a new science of resources during World War I and its uptake by various postwar technocratic movements, we can see how a set of beliefs forged in the crucible of war—particularly the belief that the flow of energy and materials was essential to national survival and that only experts could be trusted to measure and manage it properly—became pervasive in peacetime as well, giving the war a legacy that stretched far beyond the years of active combat. While many of the experts involved in wartime resource commissions lamented the paltry level of funding they received and the failure of their governments to follow their advice, the very fact that scientific expertise had been institutionalized in this way represented a major shift with repercussions for decades to come. KEPS, for example, failed to make a difference in the outcome of Russia's involvement in the war, but it laid the foundations for integrating scientific advice on industrial matters into the functioning of the Russian state, even as the nation's political structures were turned upside down. For individual scientists, regardless of whether they accomplished much of scientific value during the war itself, the experience of participating in such efforts also had significant and long-lasting consequences: it concentrated their attention on new problems, introduced them to new techniques, and sparked interdisciplinary conversations that continued and intensified in the following years.

TRACKING FLOWS

In the years following World War 1, scientists who had been involved in wartime efforts to identify and exploit strategic materials turned their attention to the processes that determined when and where such resources were "produced" and "consumed" by nature itself. In order to do so, they developed new techniques for tracking specific elements as they traveled through Earth's crust, oceans, and atmosphere and for mathematically modeling the chemical and biological processes they participated in, including the reproduction and metabolism of living beings. These techniques built on the interdisciplinary work of expert commissions established during the war to assess stocks of strategic materials, while also extending those techniques to account for change over time and to include materials and processes that, even if not directly relevant to industrial production, could nonetheless be analyzed in similar ways. As the enmities of the war faded, relationships among like-minded scientists from different nations were reestablished, with the result that innovations in the United States, Norway, Italy, and the Soviet Union were quickly adopted by scientists in other countries. By the 1930s, scientists in all these countries, working in disciplines as disparate as geochemistry, ecology, and demography, were developing sophisticated quantitative models of flows of energy and materials between living beings and their surroundings.

The trajectory of Vernadskii's research during and after the war offers a clear example of how scientists intimately involved with wartime research on strategic materials such as tungsten, aluminum, and tin could adapt the techniques they had developed during the war to the study of nature as a whole, including living beings. After Russia withdrew from World War I in 1917 and a civil war broke out that eventually led to the establishment of the Soviet Union, Vernadskii fled to Ukraine, where he shifted his focus from inventorying Russia's strategic materials to understanding the circulation of matter and energy throughout the globe, expanding a longstanding interest in the chemical and biological processes in soils to the planetary scale.[50] He focused much of his attention on measuring and comparing the elemental composition of various species—a project that he hoped would generate insights into both biology and geology.[51] This new line of research was interrupted by the chaotic conditions prevailing at the end of World War I and its immediate aftermath, which first pushed Vernadskii from Ukraine to Crimea, then briefly back to Saint Petersburg, and subsequently to Paris.[52] Only in 1926 did he finally return permanently to Russia. For the next decade and a half, he focused his efforts on studying the cycling of materials in the biosphere at the Biogeochemical Laboratory that he ran with the assistance of his former student Alexander Pavlovich Vinogradov.[53]

A similar turn from strategic materials to nature as a whole can be seen in Goldschmidt's work in Norway. After the war, he broadened the focus of the Raw Materials Laboratory from strategic materials to the task of determining "the general laws and principles which underlie the frequency and distribution of the various elements in nature," which he described as "the basic problem of geochemistry."[54] Doing so required linking the properties of atoms to the history of the planet. For this purpose, the newly available technique of x-ray spectroscopy was critical, giving Goldschmidt a means of precisely characterizing the elemental composition of minerals, including trace elements that would have been difficult if not impossible to detect using conventional chemical methods.[55] With this data in hand, Goldschmidt was able to reconstruct what he called the "metabolism of the Earth"—that is, the chemical and geological transformations that had produced the observable distribution of minerals, including the dynamic cycling of nitrogen and other elements between land, atmosphere, and oceans.[56] By the early 1930s, encouraged by a long-running correspondence with Vernadskii, he had expanded his work to include the role of living beings in reshaping the geochemistry of the planet.[57]

Even those who were not directly involved in wartime resource commissions sometimes shifted the focus of their research in response to the war's resource crises and the concerns about national survival they evoked. It was Germany's mobilization of its resources for war, for example, that first inspired the German Estonian biologist Jakob von Uexküll to expand his theory of the *Umwelt*—the "surrounding world" of perception and action of an individual organism—to entire nation-states. A German patriot, Uexküll spent the war years rallying the residents of the area surrounding his wife's estate in Pomerania to the war effort, while also publishing essays on the mobilization of the nation's resources.[58] Uexküll's prewar work on the *Umwelt* had focused entirely on individual organisms and their distinctive perceptual worlds, but his wartime essays expanded the theory to address the relationships between peoples (*Völker*), states, and the resources they required to survive.[59] In particular, he argued that each *Volk* was a natural, organic whole for which the state served as an artificial means of organizing and mobilizing the flows of resources it required.[60] In the years following the war, Uexküll's view of *Volk* and state as organized entities surrounded by an environment of resources increasingly shaped both his conservative politics and his scientific research.[61]

The growing interest in exchanges of matter and energy between living beings and their surroundings also spurred new research on rates of metabolism and growth in individuals and populations. In the 1920s, for example, Vernadskii began to emphasize the importance of what he called the "velocity" or "speed" of life in biogeochemical processes—that is, the rate at which living organisms incorporated elements from their surroundings into their bodies or transformed them before excreting them as waste.[62] While Vernadskii's work was mostly theoretical, other researchers working around the same time—all directly or indirectly influenced by the war's resource crises—developed quantitative techniques for empirically estimating life's velocity. In Italy, for example, the mathematician Vito Volterra developed a model of predator-prey relationships that was inspired by observations of the effect of wartime disruptions to fishing on the relative numbers of various species in the Mediterranean.[63] A similar model was proposed in the United States by Alfred Lotka, who drew on mathematics, physics, and chemistry to develop an approach to the study of life that he called "physical biology."[64] Like Volterra, Lotka had an indirect connection to the war: in the early 1920s, he worked in the laboratory of the demographer Raymond Pearl at Johns Hopkins University, who had become interested in modeling population growth as a result of his wartime work with the US Food Administration.[65] Back in the Soviet Union, the biologist Georgy Gause wove these strands of research together to develop laboratory models of competing populations—a project that he justified in Vernadskiian terms as "one of the ways of extending our knowledge of the distribution of the organic matter in the biosphere."[66]

Scientists also used the techniques for measuring stocks and tracking flows that they had developed during and after World War I to study how living and nonliving entities became bound together into organized systems through flows of matter and energy. The Russian zoologist Vladimir Vladimirovich Stanchinskii, for example, was inspired by Vernadskii's work to develop new instruments and methods for measuring flows of materials and energy through ecological communities.[67] Beginning in 1927, Stanchinskii carried out most of his research in Askania-Nova, a nature reserve on the Ukrainian steppe. His aim was to track the fate of energy captured directly from the sun by plants as it dissipated back into the environment or was appropriated by insects and other organisms within a given ecological community. To do so, Stanchinskii and his students developed a variety of new quantitative techniques and instruments, including deceptively simple traps that made it possible to determine the number, type, and weight of insects and other small animals within a given area.[68] The data thus collected served as the basis of mathematical models that explained why energy flowed through the system at certain rates, and why the organization of a particular ecological community—its variety of species, their relative abundance, and their relationships to one another—persisted in recognizable patterns over time.[69]

The idea of studying life by measuring the flows of energy and materials that linked living beings to their surroundings was also pursued in the United States by the Yale University ecologist G. Evelyn Hutchinson and his students. Trained in limnology (the study of lakes) and influenced by his father's work in mineralogy, Hutchinson learned about Vernadskii's work from several expatriate Russians at Yale, including Vernadskii's son, the historian George Vernadsky.[70] In the 1930s, Hutchinson launched a series of biogeochemical studies of what he called "the chemical factors of the environment that operate on living organisms," which included variations in the presence of elements across plant species, as well as the circulation of chemicals between organisms and their surroundings.[71] He also introduced a series of students and collaborators to the biogeochemical techniques and theories of Vernadskii and Goldschmidt. In the late 1930s and early '40s, for example, he encouraged Raymond Lindeman to use biogeochemical methods to study the transfer of energy and materials between living organisms and the nonliving components of their surroundings at Lindeman's research site, a senescent lake in Minnesota. In an influential paper published in 1942, Lindeman showed how precise measurements of changes in the lake's biomass and species composition could be used to reveal the flows of materials and energy between "producers" that captured energy from the sun and "consumers" that obtained energy from producers or other consumers, as well as between both producers and consumers and their surroundings.[72] Taken together, Lindeman argued, these relationships

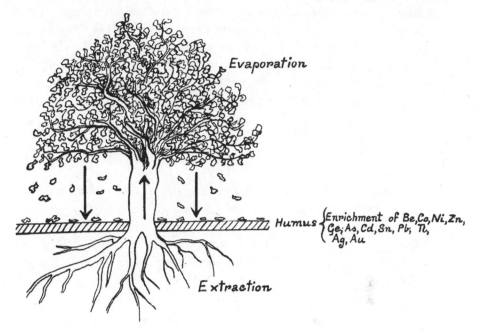

FIGURE 9. A schematic representation of the role of trees in concentrating chemical elements in the upper layers of the soil, as depicted by Victor Moritz Goldschmidt in 1937. (Reprinted from figure 3 in V. M. Goldschmidt, "The Principles of Distribution of Chemical Elements in Minerals and Rocks," *Journal of the Chemical Society* [1937]: 655–73, on 670, with permission from the Royal Society of Chemistry, conveyed through Copyright Clearance Center.)

constituted a single unit, the "ecosystem," whose functioning was the proper subject of ecology.[73]

By following the legacies of the resource crises of World War I into the interwar period and beyond, we can see how economic, military, and political concerns continued to affect the practice of science even after the conditions that gave rise to institutions such as KEPS in Russia or the Raw Materials Laboratory in Norway had passed. Vernadskii and Goldschmidt, for example, remained interested in identifying exploitable mineral deposits within their respective countries even as their focus broadened in the postwar years to encompass highly general models of biogeochemical processes and of the evolution of Earth and its life-forms. More fundamentally, they continued to use the same quantitative techniques for tracking and quantifying biogeochemical flows of energy and materials. Even Hutchinson and his students, who had not been involved in the wartime resource commissions and whose work did not have similarly obvious links to industrial concerns, claimed that their new science of ecosystems would make it possible to manage the planet's living resources more effectively, including minimizing the adverse effects of nuclear technologies developed during and after World War II on agriculture, forestry, fisheries, and human health. In these ways, we can see how the particular form of the concept of environment that was pursued through biogeochemistry and ecosystem ecology from the 1920s to the 1950s continued to bear the mark of its wartime origins.

SYSTEMS IN ENVIRONMENTS

Scientists involved in addressing the resource shortages of World War I rarely described the phenomena they were studying in terms of "environments," even though both the concept and the word itself were by then well established. Nonetheless, the data that had been collected and the techniques that had been developed to study industrial economies and strategic materials helped transform scientists' understanding of the environment following the war. In particular, by making it clear that a nation's ability to produce goods essential to its survival depended on flows of matter and energy under particular social and economic conditions, these data and techniques served as the foundation for a new way of understanding how living beings related to their surroundings. Scientists working in a range of disciplines began to argue that living beings were best understood as components of "systems," which were characterized by enduring patterns of relationships among diverse parts, each of which contributed to the survival of the whole under a particular set of external conditions. Unlike a "community," they argued, a system could not be understood by focusing solely on the relationships among the diverse living beings within it, as the majority of ecologists had hitherto sought to do. Rather, it required scientists to broaden their view to include both living beings and their nonliving surroundings, as well as the self-reinforcing circuits of matter and energy that bound them together. These ecological systems, or "ecosystems," became the focus of a new approach to ecology that was developed between the 1920s and '50s.[74]

The postwar interest in systems reflected a major shift in the theoretical framework of ecology, which had emerged in the late nineteenth and early twentieth century as a field focused explicitly on the relationship between organisms (or communities of organisms) and their environments.[75] At the foundation of this field was the recognition that there were stable relationships of interdependence among organisms that lived under given sets of external conditions, and that these relationships were both complex enough and regular enough to serve as the object of a scientific discipline. In 1877, for example, in the context of a study of oyster farming on Germany's Baltic coast, the biologist Karl Möbius described the various species necessary for oysters to thrive as a "living community" (or "biocoenosis").[76] The living community, he argued, consisted of an enduring, organized relationship among species existing under certain "external living relations" and "conditioning factors" in a given place.[77] By the beginning of the twentieth century, the living community as Möbius defined it had become the central object of research in the emerging discipline of ecology.[78] In the United States, for example, the plant ecologist Frederic Clements developed a theory of ecological succession that explained changes in plant communities in terms of shifting relationships among species in response to climate and other physical conditions. Central to Clements's theory was the notion that each living community could be considered as a unified whole—perhaps, he wrote, even itself a "complex organism, which possesses functions and structures, and passes through a cycle of development similar to that of the plant."[79]

Ecologists who embraced the concept of "system" were as interested as community ecologists had been in explaining the persistence of particular assemblages of plants and animals, but their characterization of those assemblages in terms of flows of energy and materials challenged Clements's conviction that the "living community" as a "complex organism" should be the focus of ecological research. Vernadskii, for example, questioned the primacy of both organisms and communities in ecological theory, instead arguing that "living matter" was the more fundamental unit of analysis.[80] As he defined it, "living matter" did not consist of special organic particles that gave living beings their unique properties (as Buffon had theorized in the mid-eighteenth century) but of ordinary matter that had been incorporated into the bodies of living beings, where its chemical transformations were catalyzed in ways unique to life (a view that Vernadskii based partly on Cuvier's image of life as a whirlpool or vortex of particles).[81] Stanchinskii, Hutchinson, and others were directly inspired by this view of life on Earth. Rather than carrying out studies of the mutual interdependence of living beings in a particular place, they concluded, they ought to be conducting quantitative, biogeochemical studies of flows of matter and energy between living beings and their nonliving surroundings.

Because such flows had the potential to extend far beyond any local assemblage of plants, animals, and other organisms, the biogeochemical approach to ecology also raised questions about the proper scale of ecological analysis. For Vernadskii, the most important scale was undoubtedly the planet as a whole. If all living beings taken together could be considered in terms of "living matter," he argued, then even such basic organismic functions as nutrition and respiration needed to be seen "not solely as phenomena of the organism, but also as phenomena of the terrestrial globe."[82] It was through this idea of planetary-scale metabolism that Vernadskii arrived at the idea of the "biosphere," which framed his research from the early 1920s onward. Borrowing the term from the Swiss geologist Eduard Suess, who had used it to denote the thin layer of living beings at the surface of Earth, Vernadskii redefined the term to mean the entire expanse of Earth that had been reshaped by the activity of living matter, considered as a single, highly complex chemically reactive mass.[83] This dynamic set of exchanges of energy and matter was neither arbitrary nor random, Vernadskii argued, but rather constituted a dynamic global system that was functionally organized in relation to the "cosmic milieu" that surrounded it, particularly the energy transmitted from the sun to Earth in the form of electromagnetic radiation.[84]

While Hutchinson and other ecologists in the United States took inspiration from Vernadskii's idea of the biosphere, the concept had only limited utility for those working on a smaller scale than that of the planet as a whole. However helpful it might be for understanding the history and future of the planet, it was difficult to apply to the ecology of a lake or to an assemblage of plants in a particular place. For work on these smaller scales, ecologists adopted another concept popularized by the British ecologist Arthur Tansley in 1935: the "ecosystem."[85] For Tansley, the ecosystem was an alternative to the concept of community as it had been defined by Clements and his supporters. That concept was a misleading one, Tansley believed, in that it implied a certain amount of similarity and solidarity among all its "members" that he did not believe applied to the diverse kinds of plants, animals, fungi, and bacteria and the often antagonistic relationships among them that made up ecological assemblages.[86] The ecosystem concept, by contrast, made no assumptions about the character of the entities that constituted it. Moreover, while an ecosystem was a coherent entity organized in relation to external conditions—just like a community or an individual organism—it was not limited to living beings alone. Instead, systems consisted of relationships among both living beings and nonliving entities, such as bodies of water or sediment deposits. Even if "biomes" consisting solely of the living components of an ecosystem could be singled out for analysis, that analysis would succeed only to the extent that it accounted for the system as a whole. As Tansley argued, "Various 'biomes,' the whole webs of life adjusted to particular complexes of environmental factors, are real 'wholes,' often highly integrated wholes"—but rather than being complete unto themselves, those wholes were "the living nuclei of *systems* in the sense of the physicist."[87]

Tansley himself did little to further develop the ecosystem concept, but it proved well suited to the kind of biogeochemical studies that Hutchinson, Lindeman, and others were beginning to pursue in the 1930s and '40s. In Lindeman's 1942 paper, for example, the idea of the ecosystem as a set of exchanges of energy and matter between organisms and nonliving entities was central. The continuous cycling of nutrients between living and nonliving entities in the lake he studied made the conventional distinction between a living community and its nonliving environment that had been postulated by Clements and other ecologists seem "arbitrary and unnatural," he wrote.[88] Instead, it was best to approach living beings and nonliving entities as aspects of a single "ecosystem"—that is, a "system composed of physical-chemical-biological processes active within a space-time unit of any magnitude."[89] After Lindeman's untimely death in 1942, Hutchinson and his students continued to argue for replacing the concept of community in ecology with the concept of system.[90] In the late 1940s, in particular, they joined forces with an interdisciplinary group of scientists developing a new approach called "cybernetics," which sought to understand the emergence of goal-directed behavior through mechanisms of self-regulation and self-reproduction.[91] At a cybernetics conference on the theme of "circular causal systems" in 1946, Hutchinson presented a paper that laid out the case for considering "systems" to be the fundamental unit of ecology, applicable both to groups of organisms and to the environments that they faced.[92] Since any system that was incapable of adapting to changes in its environment would be quickly replaced by a more stable system (or systems), Hutchinson argued, all the systems observable in nature were by necessity self-regulating and self-correcting. The purpose of ecology was to determine the mechanisms that made them so.[93]

If "system" was to replace "community" as the core concept of ecology, the definition of the word *environment* would necessarily have to shift along with it. It could no longer be defined as the physical conditions that shaped the life of an organism or living community, since both the system and its environment consisted of living and nonliving components. Rather, the environment was reconceptualized as the set of external conditions that influenced the functional organization of the system, whatever it might be. Both the muddy sediment at the bottom of a lake and the local atmospheric conditions that determined how quickly water evaporated from the lake's surface were nonliving, for example, but the former was an integral part of the ecosystem while the latter was part of that system's environment. Whether or not something could be considered to be part of the system depended both on the character of its relationship with other components and on the spatial and temporal scales of analysis. As those scales shifted, so did the boundaries of the ecosystem. At the scale of the biosphere itself, Earth's climate was part of the system, which in turn was surrounded by a cosmic milieu consisting of solar radiation, meteorite strikes, and the gravitational pull of the sun and the moon.

By focusing scientists' attention on the quantitative measurement of flows of energy and matter among living and nonliving entities, the concept of "system" also offered new ways of thinking about the politics of ecology and resource management. In particular, it suggested that ecological "systems" could and should be reengineered by experts in ways that "communities" could not and should not be. Vernadskii, for example, saw the recognition of the biosphere as a functionally integrated planetary system as the first step in building a progressive, liberal future for humanity as a whole—one in which the very kinds of war that had given rise to the concept of the biosphere idea would be rendered obsolete.[94] Over time, he suggested, humanity's self-awareness had grown to the point that it was now converging into a single global planetary consciousness, a *noösphere*. "The noösphere is a new geological phenomenon on our planet," he wrote, in which humanity "becomes a large-scale geological force" for the first time.[95] By refining billions of tons of pure aluminum and iron, for example, human ingenuity and industry had already begun to enrich Earth with new "biogenic 'cultural' minerals."[96] With such world-shaping powers in its grasp, Vernadskii concluded, humanity's challenge was now to figure out how to work collectively toward "the reconstruction of the biosphere in the interests of freely thinking humanity as a single totality."[97] Ecosystem ecologists were more modest in their ambitions, but they too embraced a kind of technocratic optimism that was ultimately rooted in the idea of the ecosystem as an assemblage of interdependent living and nonliving components that could be improved through engineering.[98]

For scientists who did not share Vernadskii's liberal internationalism or the technocratic enthusiasm of Hutchinson and Odum, quite different lessons could be drawn from the resource crises of World War I and the scientific and technological innovations of the interwar period. Uexküll is a case in point. Amid the political turmoil of interwar Germany, he offered a model of the "biology of the state" that contrasted starkly with the visions of planetary consciousness and technocratic management associated with the concepts of biosphere and ecosystem during the same period.[99] Rather than seeing scientists as working within a democratic context for the benefit of humanity as a whole, he saw the ideal nation-state as an organism structured by rigid social hierarchies, centered on the personal *Umwelt* of a monarchical leader, and defended against internal and external "parasites."[100] Rooted in the same wartime concerns that had motivated other scientists to search for resources necessary for victory, Uexküll offered an illiberal version of the concept of environment according to which the status of the nation as a functionally integrated, organic whole could be used to justify a ruthless war against internal and external enemies. While Uexküll did not wholeheartedly support the National Socialist government that took power in 1933, at least some Nazi Party members saw his scaled-up theory of the *Umwelt* as a useful framework for justifying both genocide and territorial expansion.[101]

The adoption of the concept of system by ecologists from the 1930s onward did not entail an abandonment of the concept of environment but rather its reconfiguration in a new semantic context. Systems were by definition not all-encompassing; they were, on the contrary, well-defined and internally organized collections of entities that were bound together in some recognizable and persistent way through exchanges of matter and energy. In this sense they were much like organisms (albeit not necessarily living), and like organisms their internal structure could be understood only in relation to the external conditions they faced. By shifting focus from organisms and communities to ecosystems, ecologists therefore did not lose their interest in or need for the concept of environment; rather, they began to think of environments in terms of quantifiable flows of energy and matter that provided the conditions for the persistence of "systems" consisting of both living and nonliving components. Behind even the radically different politics of a liberal internationalist like Vernadskii and a conservative nationalist like Uexküll, we can therefore see a set of shared assumptions and techniques rooted in the challenge of wartime resource management.

THE IRRADIATED ENVIRONMENT

With the outbreak of World War II in 1939, concerns about resource shortages reemerged as an issue of central concern for national governments. Just as in the previous world war, both combatant and neutral nations organized expert committees or reinvigorated those that had been established a quarter-century earlier, such as KEPS in the Soviet Union or the National Research Council in the United States. Scientists once again set themselves the tasks of calculating how much of each resource was needed, how much was available, and whether substitutes could be devised, and of urging national governments to invest in scientific and technological expertise. Again, just as in the aftermath of World War I, these activities persisted into the postwar years, inspiring new efforts to measure flows of energy and materials in relation to industrial needs. Moreover, a new generation of scientists repeated the same trajectory—from wartime resource crises to postwar attempts to rethink humanity's relationship to its environment—that had led Vernadskii to develop the concept of the biosphere in the 1920s. The US ornithologist William Vogt, for example, spent the war years studying the nitrogen- and phosphate-rich guano deposits of Peru and coordinating scientific cooperation between the United States and Latin America; after the war, he drew on these experiences to become a leading voice in the emerging environmental movement, warning about an impending global resource crisis and advocating for population control.[102]

Even certain aspects of the development of the atomic bomb showed significant continuity with earlier attempts to identify strategic materials and develop new ways of exploiting them. Early in the war, physicists' realization that the enormous amounts of energy released by atomic fission might be harnessed to make new kinds of weapons led to an international race to locate and control sources of uranium ore and to develop new processes for separating the highly fissile isotope uranium-235 from the more plentiful uranium-238. In the Soviet Union, Vernadskii and Fersman helped convince the Soviet Academy of Sciences that the exploitation of atomic energy could be the key both to military victory and to the country's postwar prosperity.[103] The Uranium Commission founded as a result ultimately contributed little to the war effort, but it helped lay the foundations for the postwar Soviet nuclear program.[104] In the United States, meanwhile, the Manhattan Project succeeded in producing the atomic bombs used to devastate the cities of Hiroshima and Nagasaki in 1945. Just as crucial to the success of the Manhattan Project as the designs of the bombs themselves was access to uranium, a new kind of strategic material that the United States initially procured from mines in Canada, the Belgian Congo, and the Colorado Plateau. The uranium thus acquired was subsequently processed at massive new industrial facilities that dwarfed those built by Germany to fix nitrogen during World War I. They included the Clinton Engineer Works at Oak Ridge, Tennessee, where uranium-235 was separated from uranium-238, and the Hanford Engineer Works in Washington State, where uranium-238 was transformed into plutonium-239 by bombarding it with neutrons in the world's first large-scale nuclear reactor.[105]

Some things did change significantly in the shift to radioactive materials as the fuel for new kinds of armament and a new means of producing energy. Unlike the refining of tungsten, aluminum, and other elements already present in Earth's crust, the transformation of uranium into plutonium—first accomplished in 1940 in a cyclotron at the University of California, Berkeley—actually created new atoms of an element that was so vanishingly rare that it had previously escaped the attention of physicists and mineralogists.[106] In this way, it went beyond even Vernadskii's vision of new "cultural" minerals produced through human industry but made out of existing atoms to include creating elements and isotopes that had never been observed in nature. In terms of their absolute mass, these new elements remained insignificant in comparison to other industrial materials. The United States, for example, produced less than 112 metric tons of plutonium in the half-century after 1944—a tiny fraction of the billions of tons of steel produced during the same period.[107] Nonetheless, because of their high levels of radioactivity, these elements and their decay products had a significance that went beyond their mass.

As these radioactive elements were released into individual ecosystems and the biosphere as a whole from the 1940s onward (sometimes intentionally for research purposes, sometimes as byproducts of nuclear weapons testing and the nuclear power industry), they provided new ways of tracking biogeochemical cycles and new reasons to worry about the consequences of humanity's domination of the planet. From the 1940s onward, for example, ecosystem ecologists in the United States began using radioactive isotopes to map and measure the quantities and rates of flow of materials through local and global biogeochemical systems with unprecedented precision. Unsurprisingly, Hutchinson was one of the earliest adopters of the technique, using radioactive tracers to study the movement of materials through lake ecosystems almost as soon as small quantities became available for scientific research in the early 1940s.[108] After the war, US scientists' access to radioactive isotopes and radiation sources expanded dramatically with the help of the Atomic Energy Commission and the Atoms for Peace program, the latter of which sought to give a peaceful face to the United States' ongoing nuclear weapons and nuclear energy programs.[109] The result was a tool for studying biogeochemical processes at a level of precision that earlier researchers could only have dreamed of.

Radioactive materials were not simply tracers of existing biogeochemical cycles, however; they were also interventions into those cycles in their own right. By the mid-1950s, scientists recognized that even in the absence of actual nuclear war, the biosphere was being transformed by the introduction of radioactive isotopes that had previously been entirely absent or present only in minuscule amounts, and that these transformations might bode ill for the future of human health. The appearance of radioactive strontium in milk as a result of atmospheric nuclear tests conducted in the 1940s and '50s proved particularly alarming. While some scientists remained sanguine about the capacity of humanity to manage these unruly new resources, others became increasingly concerned. Among them was Alexander Pavlovich Vinogradov, who had taken over the Biogeochemical Laboratory after Vernadskii's death. In the late 1950s, even as both the US and Soviet governments continued to suppress information about domestic nuclear accidents and radioactive contamination, Vinogradov joined other scientists in warning of the health risks of radioactive fallout. As early as 1954, for example, he was among the signatories of an unpublished report to the Soviet government warning of the effects of fallout from atmospheric nuclear weapons testing.[110] More publicly, as a participant in the Pugwash movement of scientists opposed to the proliferation of nuclear weapons, Vinogradov warned in 1959 that "atomic explosions are being conducted under the earth, in the stratosphere, in the troposphere, and in the ionosphere . . . without any attempt on the part of statesmen to understand the final results of such actions on their part."[111] In Vinogradov's warnings about the toxic effects of fallout, the optimism of his mentor Vernadskii was turned on its head. Rather than heralding the emergence of the noösphere as a new stage in the evolution of humanity and of Earth, the proliferation of "cultural" minerals indicated the slow poisoning of the biosphere as a whole.

The combination of the use of radioisotopes to map and quantify biogeochemical processes, the theories of feedback and control associated with cybernetics, and concerns about the potential ecological aftereffects of nuclear war proved central to the rise of one of Hutchinson's students, H. T. Odum, to a position of leadership in ecology in the United States in the decades following World War II.[112] After completing a dissertation in the late 1940s on what he called the "world strontium cycle" that followed closely in the footsteps of Vernadskii, Vinogradov, and Hutchinson, Odum made his scientific reputation by using radioisotopes to study exchanges of energy and matter between organisms and their surroundings.[113] In a study of the effects of atomic bomb testing on the Pacific atoll of Enewetak conducted with his brother Eugene in 1954, for example, he found that the animal parts of the coral reef absorbed much more radiation than their algal symbionts, which provided an important clue to the circulation of nutrients throughout the system.[114] Separately, Eugene Odum also conducted numerous studies of the Savannah River nuclear site in South Carolina, where plutonium and tritium were produced for the US nuclear weapons program. He and his collaborators used radioactive strontium, phosphorus, and other radioisotopes "as 'tags' in population studies, as aids in measuring energy flow rates in nature, and as a means of determining the movement of elements in biogeochemical cycles."[115] Over the following decades, the Odum brothers' biogeochemical and systems-theoretic vision of ecology profoundly shaped the field's development in the United States.

Modeling even the simplest ecosystems using such methods quickly outstripped the capacity of ecologists to calculate them by hand. To meet this challenge, ecologists turned to digital computers, whose development had been largely funded by the US military during and after World War II and which first became available to academic researchers in the late 1950s.[116] Although ecologists began experimenting with the use of computers as soon as they became available, the key turning point in their adoption came in 1962 with the launching of the International Biological Program (IBP), which aimed to promote and coordinate research on "the biological basis of productivity and human welfare." The US contribution to the IBP was mainly devoted to a program called Analysis of Ecosystems, which was directed by a former student of Hutchinson's named Frederick Smith.[117] With the aim of producing large-scale computerized simulations of the world's biomes, researchers involved in the program measured biomass and nutrient flow and built computer simulations of the feedback loops that governed the relationships among components within each ecosystem. The links between these efforts and earlier concerns with wartime resource crises were rarely made explicit, but the legacies of those efforts were clear in both the methods of the Analysis of Ecosystems program and the language used to described it. When Smith sought to justify the large US investment in the IBP, for example, he called it a contribution to understanding and managing humanity's impact on the "biosphere."[118]

In the adoption of biogeochemical methods and the language of the "biosphere" in ecosystem ecology, we can see both the continuation of projects and themes dating back to the period immediately following World War I and their transformation under new conditions. If the production of transuranic elements such as plutonium confirmed Vernadskii's vision of an unfolding noösphere in which human consciousness was critical to the energetic and material functioning of Earth, it also challenged his faith that the maelstrom of globalized, industrialized warfare could serve as a step on the path toward a community of free-thinking, self-reflective human beings. On the contrary, both the threat of nuclear war and the spread of radioactive wastes suggested that a human-dominated Earth might be closer to a sphere of death than a sphere of life—what some scholars have called a "thanatosphere," that is, rather than a biosphere.[119] In the hands of ecosystem ecologists after World War II, biogeochemistry became less optimistic, albeit no less ambitious. Recast in the technocratic language of "systems," it offered a set of conceptual and material tools for reengineering everything from individual ecosystems to the biosphere as a whole. We can thus see how certain techniques and theoretical claims could, when adapted to a new context, retain their utility even as their broader significance was transformed.

*

In 1943, with the outcome of World War II still unclear, Vernadskii wrote an essay situating the origins of his interest in the biosphere in his experience of the previous world war. Published in English translation at Hutchinson's request in the *American Scientist* just a few weeks after Vernadskii's death in January 1945, the essay explained that it was "in the atmosphere" of World War I that he had "approached a conception of nature, at that time forgotten and thus new for myself and for others, a geochemical and biogeochemical conception embracing both inert and living nature from the same point of view."[120] In other words, Verandskii wrote, it was as a result of the resource crises of the first global industrialized war, which had focused his attention on the global circulation of the "strategic materials" that kept industrial economies running and armies well supplied, that he had come to recognize the importance of Earth's "biosphere"—that is, a planetary entity composed of living and nonliving components organized into self-reproducing systems and fueled by energy from the sun.

Vernadskii's shift in focus as a result of the war was an experience shared by a multigenerational group of scientists who participated in wartime efforts to maintain industrial production and military capacity despite the disruptions and deprivations of the war, or who were later influenced by those who had. It included people directly involved in the search for strategic materials such as Victor Goldschmidt, whose work on Norway's Raw Materials Commission laid the foundations for his postwar studies of global geochemical cycles, as well as people like Alfred Lotka or Vito Volterra whose wartime work had been only marginally concerned with resources but who later came into productive contact with others who had been directly concerned with wartime supplies of food and other strategic materials. The war also indirectly influenced scientists such as Hutchinson, Lindeman, and Odum, who had been too young when the war broke out to be involved in such expert commissions and other war-related efforts themselves. Through their experiences, we can see how war both transformed the material and social environment and provided opportunities for scientists to conceptualize that environment in new ways.

In particular, by attending to the techniques for quantifying and modeling the stocks and flows of matter and energy that rose to prominence during and after World War I, we can see how the adoption of a new set of techniques lent plausibility to a new way of imagining environments and the entities that they surrounded in terms of "systems" that included living and nonliving components. As Vernadskii suggested, these methods approached both "inert and living nature" in similar ways—in particular, they reduced entities to their chemical and energetic composition, quantified their magnitudes and rates of change, and determined their functional role in a larger system or systems of which they were components. Sophisticated methods such as x-ray spectroscopy allowed Goldschmidt, Hutchinson, Odum, and others to quantitatively determine the chemical composition of living and nonliving bodies. Even simple techniques for counting and measuring living beings—such as those used by Stanchinskii in Askania-Nova and Lindeman in his trophic-dynamic study in Minnesota—proved remarkably generative when linked to mathematical models of the flow of energy and materials. This was particularly the case when those models could be implemented in digital computers, as they were in the decades following World War II.

One of the things that made these methods distinctive was precisely the fact that they could be applied in similar ways to both living and nonliving entities. The concepts of "biosphere" and "ecosystem" that were developed between the 1920s and the 1940s drew on the data produced by these techniques to suggest a new model of nature centered not on "organisms" or "living communities" but instead on "systems." Like an organism or a communitiy, a system was an assemblage of diverse parts that were functionally integrated and organized in relation to the conditions that surrounded them. Unlike an organism or a community, however, a system did not consist solely of living beings. Since the theorists of ecosystems assumed from the outset that those systems included nonliving components—some of them possibly human-made or at least human-influenced—they were also more open to altering or replacing those components than most community ecologists had been. Where community ecologists had seen human activities as "disturbances" of natural processes of ecological succession, ecosystem ecologists saw changes in the feedback loops of self-organizing systems. Reflecting its roots in wartime resource management, the systems view thus opened the door to technocratic and managerial approaches to solving ecological problems.

Even after the end of World War II, concerns about national security continued to shape the sciences of biogeochemistry and ecosystem ecology. Indeed, the Cold War heightened the interest of scientists and governments in methods that could be used to predict and control the flow of resources and energy through natural and artificial systems. Using radioactive tracers and computerized models, ecologists showed how complex systems of living and nonliving components were organized through feedback loops that allowed them to maintain their structure under particular external conditions. The awareness that humanity had introduced radically new and potentially harmful substances into the biosphere also raised new concerns, however, about the potential for humanity to damage or even destroy its own environment. These concerns contributed to the emergence of a self-conscious "environmental movement" in the 1960s and '70s, when consumers in the United States and elsewhere began to realize that the safety and healthiness of the products they purchased could not be disentangled from the quality of the environments in which they were produced. The next chapter turns to these consumers and to the activists who laid the foundation for a new kind of popular "environmentalism."

The Evolution of Risk: Toxicology, Consumption, and the US Environmental Movement

When applied to plants, the chemical aminotriazole (also known as amitrole) is capable of inhibiting the production of a protein essential to photosynthesis, making it a highly effective broad-spectrum herbicide. From an environmental and toxicological perspective, one of its most appealing properties is that it degrades quickly in soil and water, reducing the chance of unwanted side effects to human health. Only when it is applied to food crops immediately before or during the growing season is there a risk that the chemical might be incorporated into the plant, leaving residues that survive unchanged until they reach the consumer. When aminotriazole was first marketed as an herbicide in the United States the mid-1950s, farmers were therefore told to apply the chemical only after the growing season had ended.[1] Cranberry growers were among those who were quick to adopt the new chemical, which allowed them to easily free their bogs of poison ivy and other perennial weeds. It soon became apparent, however, that not all growers were following the recommended guidelines for its use. A few weeks before the Thanksgiving holiday of 1959—that is, at the start of the season when the vast majority of cranberries were sold in the form of fresh berries or cranberry sauce—the US government announced that traces of aminotriazole had been detected in lots of cranberries. All the cranberries harvested over the previous year would therefore be withheld from the market until they could be tested for contamination.[2]

The legal justification for the decision was the so-called Delaney Clause of the 1959 amendment to the Food, Drug, and Cosmetic Act, which prohibited the sale of any product containing measurable quantities of substances known to be carcinogenic to animals or humans. Unluckily for cranberry growers, recent studies had shown elevated rates of thyroid tumors in rats fed high doses of aminotriazole, meaning that the Food and Drug Administration was authorized to remove cranberries with even the slightest trace of the chemical from the market. In the wake of the announcement, nationwide sales of cranberries and cranberry sauce suffered a "crushing blow," in the words of one industry observer, with many stores pulling all cranberry products from their shelves regardless of the likelihood of contamination.[3] Growers were quick to defend themselves, launching a media blitz to convince consumers that cranberries remained as safe as ever and pressuring politicians to defend the industry.[4] Within a week of the government's announcement, both Republican Vice President Richard Nixon and Democratic presidential candidate John F. Kennedy had ostentatiously consumed cranberries in public to demonstrate their support.[5] Growers also took their case to the courts, arguing that those who had never used the herbicide, or who had used it properly, had been unfairly damaged by association with the very small minority who had allowed their harvest to be contaminated.[6]

These efforts to restore the good name of the cranberry industry proved largely successful. Eventually the US government agreed to indemnify up to $10 million of the industry's losses, and as early as the spring of 1960 a marketing blitz in the weeks before Easter led to record sales.[7] In subsequent years, cranberry sales not only recovered but exceeded their pre-1959 levels; in 1964, for example, total US production was nearly two and half times greater than in 1959, while the price per barrel shot up by 50 percent.[8] One reason for the rebound was that the cranberry industry diversified its product line to include sweetened juices and other processed products—a decision made partly in response to the cranberry crisis, which had revealed the industry's dependence on holiday sales of fresh berries and canned sauce. Another was that herbicides such as aminotriazole, pesticides such as DDT, synthetic nitrogen fertilizers, and mechanized harvesters and other equipment had raised the yield of cranberries per acre even as labor costs dropped. The cranberry crisis of 1959 therefore proved to be only a brief setback in the industry's steady growth.[9]

The cranberry industry and its growing product line were just one small part of the booming US economy in the postwar decades, which provided consumers with a dizzying array of new products to choose from but also generated new anxieties about the hidden hazards those products might contain. From the 1950s onward, celebrations of consumer abundance were shadowed by concerns ranging from cultural homogenization to unintentional mass poisoning. To allay these concerns, consumers turned to experts who had developed techniques for detecting residues of potentially harmful substances in consumer goods, tracking the dispersal of those substances through ecosystems and populations, and assessing the risks those substances posed to the health of humans and other living beings. For those experts, the postwar world of consumer abundance was not only a source of concern but also an opportunity to generate new kinds of knowledge and establish new kinds of expertise. The explosion in the number of new chemical substances coming onto the US market in the decades following World War II, for example, provided not only materials for new products that potentially carried with them new risks, but also a new set of tools and tracers that could be used to reveal hitherto unsuspected relationships between bodies and their surroundings.

These new concerns and techniques served as the foundation of a new articulation of the concept of environment. In the 1950s and '60s, a number of biologists, writers, and activists began to apply ideas about the relationship between organisms, communities, and ecosystems and their surroundings—including recent discoveries in genetics and evolution—to the problems of contemporary life. In particular, they argued that the flood of synthetic herbicides and other artificial substances being produced by the consumption-driven US economy was threatening the ability of living organisms of all kinds, including humans, to adapt and survive. The best known among this group is probably Rachel Carson, a science writer trained in biology whose 1962 book *Silent Spring* documented the abuse of what she called "biocides," including herbicides such as aminotriazole and pesticides such as DDT.[10] The core argument of the book was not simply that these specific chemical compounds were toxic, but that substances like them were being invented and distributed at such a prodigious rate and in such an indiscriminate fashion that neither humanity nor the web of life of which it was a part could evolve defenses against them rapidly enough to avoid serious harm. In other words, Carson argued, it was precisely the technical ingenuity that earlier environmental thinkers had celebrated that now threatened humanity's long-term survival.

Between the 1950s and 1970s, the idea that each individual was surrounded by an environment of invisible toxic risks became compelling to large swaths of the US public in a way that previous versions of the concept of environment had not. By the end of this period, what we now know as "environmentalism" had emerged—that is, a social and political movement built around the idea that humanity's growing power to manipulate the material world was unintentionally harming the very environment on which it depended for survival. The movement grew in part because it promised to unify Americans across and despite their intensifying social divisions. Carson and other environmentalists of the period argued that all Americans—indeed, all humans—were, as members of a single species, vulnerable to the same toxic risks, and that everyone therefore had a stake in ensuring a healthy environment. This idea of a single humanity surrounded by a shared environment of toxic risks also had consequences for the way environmental activists described the environment. Rather than pointing to the specific environments of particular organisms, people, communities, or ecosystems, they often spoke of *the* environment of humanity, or even of the community of life as a whole. Although this universalist environmentalism was politically effective in the 1960s and '70s, it was also deeply problematic, as the environmental justice movement in the following decades would make clear. Even as they remained focused on toxic risks to human health, a new generation of activists challenged the environmental movement's universalism by calling attention to the risks faced by those who benefited least from the United States' prosperity.

CONSUMING

In the decades after World War II, the US economy expanded rapidly on the strength of an ever-increasing rate of consumption that transformed the everyday conditions of life for many of the country's inhabitants. This acceleration of consumption had multiple causes, including the postwar baby boom; productivity gains linked to mechanization and automation; the growth of the marketing and advertising industries; and the development of new products, including pharmaceuticals, plastics, pesticides, and other synthetic chemicals. The postwar turn toward consumption was also the hoped-for result of a set of economic reforms implemented by the US government from the late 1930s onward under the influence of the economic theories of John Maynard Keynes, who had argued that the fastest way to end the Great Depression was for government to directly stimulate consumption.[11] Franklin Delano Roosevelt's New Deal government therefore put into place policies ranging from purchasing goods and services directly, to establishing public works programs, to supporting the demands of labor unions for higher wages, to increasing the supply of circulating money and facilitating consumer access to credit.[12] Despite the necessity for rationing and price controls during World War II, such consumer-centered policies continued after the Depression was over, helping to fuel the postwar consumption boom.[13]

For the typical middle-class consumer, the result of these macroeconomic policies and shifts in business and technology was access to an unprecedented diversity and abundance of goods. In the local grocery store, the big-city department store, or the diverse array of retail outlets available in the new suburban shopping malls, not only were familiar products now offered more consistently and at lower prices than they had previously been, but entirely new products appeared for the first time. Consumers learned about these products in part through a dramatic expansion in advertising in traditional print media as well as in radio, television, roadside billboards, and other venues.[14] In addition to informing consumers about the range of new options available to them, postwar advertising also promised—in ways tailored toward specific demographic segments of the US population—that happier, healthier, more exciting, and more fulfilling lives were just a purchase away.[15] Some of these new products were made of familiar materials that had been transformed by novel production processes or incorporated into new designs, such as the molded plywood furniture of Charles and Ray Eames; others were made of entirely new substances that had been recently developed in chemical laboratories, including plastics, dyes, scents, detergents, pesticides, herbicides, and pharmaceuticals.[16] By 1958, for example, US manufacturers were producing 920 million pounds of polyethylene each year for everything from shampoo bottles to children's toys to garbage cans.[17]

The culture of consumption also transformed the landscapes in which Americans lived, worked, and played. Federal housing policies encouraged home ownership, municipal development plans fostered suburban sprawl, and racialized patterns of settlement—reinforced by discriminatory lending policies and restrictive neighborhood covenants—encouraged many members of the white middle class to flee the increasingly diverse inner cities in their search for racial homogeneity as well as for space, security, and other amenities.[18] Their new suburban homes gave them ample space to fill with consumer goods, while the homes themselves were major purchases that required costly maintenance, often made possible through generous government-backed loans. Suburbanization also intensified energy consumption, lengthening the distances Americans traveled to sites of work, education, consumption, and leisure while increasing the energy required to heat and cool their homes.[19] At the same time, new spaces of consumption were being built across the suburbanizing US landscape to make the purchase of consumer goods as convenient, affordable, and enticing as possible. The department stores that had served as palaces of consumption in city centers since the late nineteenth century were now complemented by suburban supermarkets, shopping malls, and fast food restaurants, all targeted toward encouraging consumption by automobile-equipped consumers in a sprawling metropolitan landscape.[20]

This new world of consumption was linked to profound changes in systems of production and distribution. In the case of agriculture, for example, the widespread postwar adoption of synthetic pesticides and fertilizers, gasoline-powered vehicles and other machines, and new crop varieties designed for vigorous growth and easy harvesting led to dramatic increases in yields, not just for specialty crops such as cranberries but for a range of staples. With the help of federal policies that provided loans, supported prices, and insured against crop failure, farmers adopted capital- and input-intensive methods that favored the consolidation of farms, the growth of agribusiness, and regional specialization. One of the consequences was that food moved longer distances and underwent more profound transformations on its journey from farm to table than it had in earlier eras. Advances in refrigeration technology, new systems of warehousing, and the rise of the independent trucker brought California vegetables, Iowa grains, and a variety of processed and frozen foods to customers in distant markets at cheaper and cheaper prices, replacing the rail-based networks that preceded them.[21] For consumers, these developments meant that a wider range of more affordable foods was available throughout the year.

As the geographical and social distance between producers and consumers increased, it became correspondingly difficult for consumers to know how the products they encountered on store shelves had been made.[22] Like consumers everywhere, Americans had traditionally assessed the quality and value of the products they consumed through a combination of methods, including direct relationships with local producers. Now it was no longer possible, in many cases, to know precisely where a product had been made, let alone to be personally acquainted with the people who had made it or the conditions under which they had done so. As a result, although these new production and distribution systems helped give middle-class Americans a sense of unprecedented prosperity and freedom, they also undermined trust, generated anxiety, and drove demand for new forms of advice and expertise. Chemical herbicides and pesticides and other products of the booming chemical industry were especially worrisome. In addition to being ubiquitous and unfamiliar, they were also associated with the arcane practices of scientists and engineers. Some Americans, concerned about the hidden risks of the consumption-driven economy and aware of their inability to assess them, sought out alternatives such as growing their own organic vegetables or following the guidelines for healthy eating and living prescribed in magazines such as J. I. Rodale's *Organic Gardening*, first published in 1942, whose subscriber base skyrocketed in the 1950s.[23]

SAVE 20¢ on 2 cans

Ocean Spray CRANBERRY SAUCE   *Ocean Spray* CRANBERRY SAUCE

Ocean Spray Cranberry Sauce

LIMITED OFFER

Buy any two 16-oz. cans of Ocean Spray Cranberry Sauce. Send both labels to Ocean Spray, P.O. Box 66, New York 46, N.Y.—and we'll send you 20 cents in the next mail. ACT NOW!

"Delicious with Chicken"

"Grand with Ham"

"Perfect with Pork Roast"

"The 'most' with Pot Roast"

Ocean Spray Cranberry Sauce
—It's the natural mate for every meat!

FIGURE 10. An advertisement for Ocean Spray cranberry sauce published in the fall of 1958. (*Atlanta Constitution*, October 9, 1958, 16.)

By attending to how a system for generating material abundance could simultaneously produce feelings of prosperity and insecurity, we can see how a particular state of affairs and the meanings that were attributed it emerged together in a particular time and place. The landscape of consumption in the postwar United States consisted both of its material components—that is, the consumer goods themselves, along with their supporting infrastructures of production, distribution, and disposal—and of the attitudes and values that Americans brought to it. Anxiety about the safety of one's food was neither simply a direct reaction to a changing food system nor simply a projection of unrelated anxieties (about the possibility of nuclear annihilation, for example) onto any arbitrary object that happened to be at hand. Rather, as the food system changed in ways both intended and unintended—presenting consumers with a range of unfamiliar products and new modes of consumption—consumers sought to respond to that system in ways that fulfilled their own aims and met their own expectations. Their hopes and anxieties about this new system were therefore reducible neither to the physical properties of the goods nor to the values and attitudes they brought to them; they emerged, rather, out of the encounter of the two.

DETECTING

At the very same time that Americans' concerns about the hidden risks of consumer goods were growing, scientists were developing new methods of detecting minuscule traces of potentially harmful substances and assessing the risks those substances posed to the health of humans and animals. From the toxicologists, ecologists, and epidemiologists who carried out such studies, consumers learned that egregious forms of contamination that produced immediate harm were not the only things they had to fear. Toxicity could also be subtle, diffuse, and slow acting, producing life-threatening illnesses such as cancer over the course of numerous individual exposures, each of which on its own might seem insignificant. In the case of aminotriazole, for instance, neither growers nor consumers were able to detect the trace residues that remained on affected berries without the help of chemical analysis, and the studies that revealed its potential carcinogenicity suggested that it would have an effect only after years of daily exposure.[24] As the sensitivity of techniques for chemical detection improved, moreover, it became clear that synthetic chemicals were present not only in contaminated foods but also in ambient air, water, and soil, revealing a world of hidden toxic risks.

The techniques that revealed these risks had their roots in the fields of industrial hygiene and toxicology as they had developed in the first half of the twentieth century. Well into the 1930s, however, rather than studying consumers exposed to low doses of toxic substances, practitioners in these fields focused their efforts on workers and on the acute effects of large exposures. As a rule, they believed that companies that produced or employed potentially toxic chemicals needed to take proper precautions to protect their workers, but that these companies had no special responsibilities to the products' consumers or to the people who lived near sites where the products were made, used, or disposed of, beyond warning them of any acute toxicity. In the 1920s and '30s, for example, Robert Kehoe conducted studies of lead exposure at plants that produced the gasoline additive tetraethyl lead.[25] Like many others of the time, he assumed that there was a level of exposure below which workers would suffer no ill effects and which consumers would almost certainly never exceed.[26] Moreover, when he looked beyond the laboratory through studies of lead levels in blood samples taken from Americans living in industrial landscapes and indigenous Mexicans living under what he described as "primitive conditions," he concluded that lead exposure was both normal and natural, and therefore that the introduction of leaded gasoline would have little impact on human health.[27] Even as he was stressing the responsibility of manufacturers to protect their workers from daily exposures to lead, Kehoe was also absolving them of responsibility for what happened beyond the factory walls.

To the extent that industrial hygienists and toxicologists before World War II looked at toxic exposures beyond the workplace, then, it was mainly with the aim of demonstrating that such exposures were essentially harmless. It was only in the years following the war—as the number of synthetic chemicals sold and used in the United States exploded and as consumer anxieties grew—that toxicologists and industrial hygienists began to seriously consider the cumulative impact of numerous nonoccupational exposures to very small doses of potential toxins. Even Kehoe, whose research in the 1930s had been used by the companies he advised as ammunition in their battle against government regulation, argued for broadening the purview of industrial hygiene to address toxic risks faced by consumers and citizens in their environments beyond the factory. As he noted at a 1958 conference in Washington, DC, on the theme of "Man and Environment," the "gap between technology and biology" had widened in recent decades, with the result that humanity was now much more capable of transforming its environment than it was of understanding the consequences of that transformation for its own health.[28] Industrial hygienists therefore needed to broaden the scope of their research to take these new capacities and risks into account.

An important technical factor in the shift from the factory to the consumer environment was the development of new techniques for rapidly and precisely detecting tiny amounts of specific chemicals, including complex organic compounds such as aminotriazole or DDT.[29] In the postwar decades, among the most important of these techniques were paper and gas chromatography. Paper chromatography was first developed in the mid-1940s and came into wide use over the following two decades. Simpler and speedier than conventional methods of chemical analysis, it relied on the fact that compounds present on a strip of filter paper would be carried at different rates by a solvent as it traveled from one end of the strip to the other.[30] Other chemical reactions were then used to give the substance of interest a distinctive color, allowing it to be identified and roughly quantified. A more complex but also more sensitive and precise variant of the technique, gas chromatography, became available in the 1960s. It was capable of detecting aminotriazole, DDT, and other pesticides at minuscule concentrations, down to less than a single part per million. When cranberry lots were being tested for aminotriazole in 1959, the Food and Drug Administration was still using conventional methods of chemical analysis, but by the 1960s the introduction of chromatography and other techniques had made it possible to rapidly test large quantities of goods for a range of potential chemical contaminants.[31]

Determining whether these synthetic chemicals were actually harmful to humans or other forms of life at typical levels of exposure was a different question, one that required shifting from chemistry to toxicology. However, for both practical and ethical reasons, directly assessing toxicity in humans was challenging, particularly when researchers were interested in long-term as well as acute effects. Some toxicologists and physicians did conduct experiments with human subjects, albeit usually with doses that were presumed to be unlikely to cause serious injury or death, but under most circumstances human experimentation was limited by the obvious risks entailed by experiments with substances that were used precisely because they were toxic to some forms of life.[32] Accidental exposure occurring during the normal course of production, transportation, or use of agricultural or industrial chemicals provided some opportunity to study toxicity without running afoul of such concerns.[33] Such occupational exposures were very unlike those experienced by consumers, however, who were generally exposed to small quantities of a wide variety of chemicals rather than a massive dose of a single chemical.

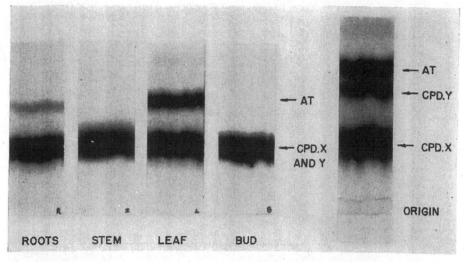

FIGURE 11. A chromatogram showing the presence of aminotriazole (AT) and two unknown derivatives (CPD.X and CPD.Y) in the roots, stems, leaves, and buds of plants. (Reprinted from figure 2 in David Racusen, "The Metabolism and Translocation of 3-Aminotriazole in Plants," *Archives of Biochemistry and Biophysics* 74, no. 1 [1958]: 106–13, on 110, with permission from Elsevier.)

Estimates of toxicity to consumers therefore continued to rest heavily on analogies to controlled laboratory experiments with nonhuman animals, which depended on the assumption that humans and animals—usually rats, but sometimes dogs or other species—were similar enough that an exposure that was toxic to the latter would also be toxic to the former. Since the 1920s, researchers had been developing standardized laboratory tests that made it possible to quantitatively compare the toxicity of various substances. For such tests, they fed rats or other animals increasing amounts of a given substance in order to determine the "lethal dose" at which half of the animals died, the so-called LD50.[34] By making certain assumptions about the shape of the dose-response curve and the relationship between animal and human responses, toxicologists could derive a "threshold limit value" or "maximum allowable concentration" to which workers and consumers could be safely exposed.[35] Suitably modified, such techniques could also be used to study the long-term effects of exposure. In an early study of aminotriazole that seemed to definitively demonstrate its carcinogenicity, for instance, toxicologists at American Cyanamid exposed rats to drinking water containing aminotriazole in concentrations of up to 100 parts per million over the course of two years. When they killed and dissected the rats, they found that those who had received the highest doses had elevated rates of thyroid tumors.[36]

Even when seemingly clear evidence of toxicity emerged from laboratory animal studies, however, extrapolating from the results of continuous high-dose exposure in rats or dogs to occasional low-dose exposure in humans proved controversial. In the paper announcing the results of their study of aminotriazole exposure in rats, for example, American Cyanamid toxicologists Thomas Jukes and Boyd Schaffer argued that the findings actually pointed to the safety of the herbicide. For one thing, the fact that daily exposure to large amounts of a chemical over long periods of time increased the likelihood of cancer in rats did not necessarily mean that occasional exposure to small amounts of the same chemical would lead to cancer in humans. As Schaffer told one journalist in 1959, a human being "would have to eat 15,000 pounds of cranberries a day for many years" to mirror the experimental conditions used in the study.[37] "If you ask me whether the very small amount of residue present is dangerous to people," he continued, "I steadfastly maintain the answer is 'no.'"[38] Moreover, Jukes and Schaffer claimed, the mode of action by which aminotriazole led to tumors was shared by antithyroid compounds found naturally in a wide range of foods, from milk to cabbage. At the doses to which consumers were likely to be exposed, Jukes later argued, the harm was likely to be lower than the cumulative effect of consuming foods that were already part of "everyone's daily diet."[39]

Even as techniques of toxicological testing improved, uncertainties about the tests' relevance to human health continued to be raised by toxicologists themselves. In the mid-1970s, for example, a new kind of assay was introduced that promised to determine the carcinogenicity of a substance without requiring laborious and time-consuming tests with living animals. Named for its inventor Bruce Ames, the so-called Ames test was based on the assumption that cancers are caused by mutations in a cell's genetic code that impair the normal processes of cell growth and reproduction—that is, that "carcinogens are mutagens," as Ames and his colleagues titled a 1973 paper.[40] This claim opened the door to new tests for carcinogenicity based not on the growth of tumors, which was often slow and dependent on many factors besides chemical exposure, but on the detection of chemical mutations to the genetic code. By accelerating the screening of potential carcinogens used in flame retardants, hair dyes, and other consumer products, the Ames test helped to dramatically expand the realm of environmental risk.[41] As its use was broadened to include myriad other substances to which consumers were exposed, however, it became clear that synthetic chemicals were not unique in their capacity to cause genetic mutations. Echoing Jukes and Schaffer's earlier claims about aminotriazole, Ames concluded that humans and other organisms had little to fear from synthetic toxins, since their evolutionary history had granted them defenses against a world full of "natural mutagens, teratogens, and carcinogens."[42]

Whether laboratory studies exposed animal or human subjects to potentially toxic chemicals or used bacteria to test them for carcinogenicity, they could tell scientists only about the effects and mechanisms of varying doses of a given substance under controlled conditions. The extent to which people, plants, or animals were actually exposed to toxic levels of that substance in the real world was yet another question, one that could be answered only by looking outside of the laboratory. When ecologists found an elevated number of sick or dying fish following the treatment of a lake or its surroundings with a pesticide or herbicide, for example, they could conclude that the chemical was causing harm, even if they remained ignorant of the details of its mode of action and the precise threshold of toxicity for various forms of life. In 1945, the US Bureau of Entomology and Plant Quarantine and the US Fish and Wildlife Service conducted a joint study of the effects of DDT spraying on wildlife, finding that even a single spraying could harm populations of animals other than the mosquitoes, fire ants, gypsy moths, and other insects that were its intended targets.[43] Other studies with similar results soon followed, raising concerns that spread beyond insects and birds to include human populations.

The development of improved techniques for detecting toxic substances and determining the risks they posed to human and nonhuman forms of life did not, by itself, carry any necessary implications for what should be done about the production and distribution of such substances. Indeed, certain complexities and uncertainties not only persisted but actually increased as these techniques were improved, as the expanding use of the Ames test reveals. Advances in chemistry and toxicology thus exacerbated anxieties about the world of consumer abundance, even as they sought to respond to these anxieties. In this limited sense, the critics of the Delaney Clause and the decision to remove cranberries from the US market in 1959 were correct: although the ultimate aim of such measures was to improve the safety of the US food system, the immediate effect was to reveal the all-pervasive and unavoidable nature of environmental risks and thereby exacerbate the very fears they were intended to allay. In this case as in others where technical advances have opened up new perspectives on the environment, we can see how such advances often are more effective at redefining the problem than providing solutions.

REGULATING

As the number and diversity of potentially toxic substances to which consumers were exposed increased, along with the sensitivity of methods capable of detecting chemicals and assessing their potential for harm, Americans turned to the federal government for assurance that they products they consumed were safe. At the time, however, consumer protection in the United States continued to be based on a model institutionalized in 1906 by the Pure Food and Drug Act, which had created the Food and Drug Administration (FDA). The law was based on several implicit premises: that giving consumers accurate information about the products available for purchase was sufficient to allow them to protect themselves from risk, that the greatest hazard came from the hidden presence of acutely toxic substances, and that risks to those who might incidentally come into contact with the toxins in question were both rare and beyond the scope of the government's authority. The same premises shaped the Food, Drug, and Cosmetic Act (FDCA) of 1938, which strengthened the FDA's powers of enforcement in response to a series of contamination and adulteration scandals.[44] They also influenced the 1947 Federal Insecticide, Fungicide, and Rodenticide Act, which tightened the regulation of pesticides, herbicides, and other toxic substances with applications in agriculture or other industries.[45] While these laws granted the government new authority to regulate the labeling and marketing of toxic substances, they focused solely on direct risks to the consumer, and they assumed that the government's responsibilities ended once it had ensured that the product had been accurately represented by the manufacturer.

Over the course of the 1950s, this approach to consumer protection came under fire by politicians and activists who believed that the increasing number of novel substances in consumer goods demanded more aggressive and wide-ranging measures. One of them was James Delaney, a congressman from New York who convened a series of hearings on food safety in 1950/51 that aimed to demonstrate the need for tighter and more comprehensive regulation of "food additives," a category including any substance used in the production of a food item that was still present in the final product, such as pesticides and herbicides.[46] One of the most important witnesses at the Delaney hearings was

Wilhelm C. Hueper, who had worked as an industrial hygienist for DuPont in the mid-1930s before becoming a prominent advocate for the need to control "environmental cancer" both within and beyond the workplace.[47] Hueper's testimony influenced several people who would later go on to become important voices in the environmental movement, including Rachel Carson and Murray Bookchin.[48] While the hearings did not immediately lead to new legislation, they helped lay the foundation for the Delaney Clause of the 1958 Food Additives Amendment to the FDCA, which prohibited the sale of foods shown to contain any amount of any additive known to cause cancer in animals or humans.[49] The zero-tolerance nature of the Delaney Clause, which was first invoked in the cranberry crisis of 1959, made it one of the amendment's most controversial provisions.

Although the Delaney Clause did not explicitly mention the environment, it prepared the way for later environmental laws by expanding the definition of "food additives" that were subject to federal regulation. An herbicide such as aminotriazole could become an "additive" for the purposes of the law not because growers or processors had intended it to be present in the fruits or vegetables they sold — that is, not because it matched the commonsense definition of the term—but because detectable residues of a carcinogen had been left in the final product. As a result, the Delaney Clause implicitly extended the government's regulatory authority to include any aspect of growing, processing, or packaging that had the potential to leave similar residues. Moreover, its focus on cancer and its prohibition of even the tiniest residue of a potentially carcinogenic substance brought attention to the repeated exposure of consumers to trace amounts of substances that might result in illness over the long term. According to the plain language of the clause, the magnitude of the risk to humans at typical exposure levels was irrelevant; all that mattered was that some dose of the substance was capable of producing cancer in animals. As one newspaper account noted, the clause was not directed at toxins that caused immediate and obvious harm, which were already regulated by existing laws, but rather at those "insidious" substances "that can cause injury slowly through prolonged use."[50]

In the 1950s and '60s, consumer activists added something else that was crucial to the emergence of explicitly environmental legislation—namely, a concern for the impact of toxic chemicals on people who neither directly used them in the workplace or at home nor consumed products contaminated by them. Seeking to improve Americans' quality of life at a time of unprecedented material abundance, these activists were skeptical of claims that industry and government would improve product safety without being compelled to do so or that informing consumers of a product's risks was sufficient. Instead, they advocated for expanding on traditional consumerist concerns about product safety and truth in advertising to include the environmental consequences of consumption. This meant not only considering repeated low-level exposures to toxic substances through contaminated goods but also the effects of incidental exposures of nonconsumers. In Long Island, for instance, a group of community members filed a lawsuit against the spraying of DDT in the late 1950s to control populations of the invasive gypsy moth.[51] Although unsuccessful, their lawsuit played a key role in galvanizing environmental activism in the United States—among other things, leading to the establishment of the Environmental Defense Fund (EDF), the first major new organization to make the concept of environment central to its mission. Carefully distinguishing itself from organizations focused on protecting private property rights, EDF instead aimed to establish "a body of common law under which the general public can assert its constitutional right to a viable, minimally-degraded, environment," even when that meant constraining the freedom of manufacturers, consumers, or the government.[52]

Self-consciously "environmental" activism was therefore already underway in some places by the end of the 1950s, but it was only with the widespread media coverage of Carson's *Silent Spring* in 1962 that it came to national attention. Inspired in part by the cranberry scare, the book drew on the work of local activists as well as a wide range of toxicological and ecological studies to argue that the misuse of "biocides" was a concern for all citizens, not just for the people who purchased them and used them.[53] In 1963, a series of congressional hearings on pesticides gave the issue additional prominence. These hearings remained focused on the consumer; as Minnesota Senator Hubert Humphrey declared on the first day of the hearings, "There is one supreme interest—the health of us all. We are all, in the final analysis, consumers."[54] At the same time, witnesses such as Carson and Hueper broadened the scope of consumer protection to include not only the risks posed by the direct consumption or use of a given product but also the risks that such consumption generated for others. Pesticides, herbicides, and other toxic substances could easily make their way into the bodies of those who neither purchased nor consumed the products that contained them, they argued, but consumers were ill equipped to assess such risks, industry had consistently failed to address the risks of its own accord, and existing federal laws did not empower the FDA to intervene. Thus, they concluded, new legislation was needed to address what

Hueper called "the growing contamination of the human environment, that is the air, water, foodstuffs and consumer goods, with manmade chemical and physical agents detrimental to human health."[55]

Not unlike the earlier Delaney hearings on food additives, the congressional hearings inspired by *Silent Spring* had few immediate consequences for public policy. Nonetheless, they contributed to a gradual but significant shift in the regulation of toxic substances away from the informed-consumer model that had dominated through the 1950s. A watershed moment was the passage of the National Environmental Policy Act (NEPA) in 1969, a law largely crafted by political scientist Lynton Caldwell, who had begun arguing for the benefits of making the "environment" into "a new focus for public policy" in the early 1960s.[56] In addition to requiring environmental reviews for a wide range of federal activities, NEPA established the Environmental Protection Agency (EPA) and authorized it to regulate toxic substances both for their immediate impact on consumers, as the FDA had done since the beginning of the twentieth century, and for their broader environmental impact. Following the passage of NEPA, explicit references to "the environment" became increasingly common in new legislation, including the 1972 Federal Environmental Pesticide Control Act, which gave the EPA the power to prohibit pesticides that had "unreasonable adverse effects on the environment," and the 1976 Toxic Substances Control Act, which bolstered the EPA's ability to regulate chemicals that posed "an unreasonable risk of injury to health or the environment."[57]

By considering these laws as expressions not simply of a new concern with the state of the environment but also of a new conception of what the environment was, we can better understand both their strengths and their limitations. Building on toxicological discoveries about the ability of toxic substances to circulate far beyond the bodies of the workers who produced them or the consumers who bought them, they shifted the focus of the federal regulation of pesticides from informing consumer choices, long the basis of consumer protection in the United States, to safeguarding the healthiness of the environment within which those choices were made.[58] In this way, these laws also contributed to a new vision of the American consumer—namely, one whose health and safety was recognized as being dependent on his or her own informed choices, distant production systems, the choices of other Americans, and the landscapes they shared. At the same time, by focusing on consumers and their exposures to hidden chemical risks, these laws drew attention away from other kinds of subjects and other kinds of threats.

THE UNIVERSAL ENVIRONMENT

The idea that humanity might not be able to adapt to the environment it had created for itself was one of the conceptual foundations for the environmental movement that emerged between the 1950s and the 1970s. While postwar environmentalists were hardly the first to argue that progress had brought new and unexpected problems, the specific way in which they did so was distinctive.[59] In particular, they were strongly influenced by developments in evolutionary theory and population genetics that became known as the "modern synthesis" after the publication of *Evolution: The Modern Synthesis* by the British biologist Julian Huxley in 1942.[60] Between the 1920s and '40s, following decades of growing doubt about the importance of Darwinian natural selection to evolution, biologists such as Sewall Wright, Theodosius Dobzhansky, Ernst Mayr, and George Gaylord Simpson had revitalized evolutionary theory by showing how the traits of a population could shift over time as the result of random genetic mutations and selective pressures. Central to the modern synthesis was the idea that species evolved not because organisms strove to adapt themselves to their environments in pursuit of some inner purpose or teleology, as evolutionists from Lamarck onward had argued, but rather because those species that failed to adapt to changes in their environment were ruthlessly eliminated, along with the genetic variations that had undermined their evolutionary fitness. As Dobzhansky noted in his 1937 book *Genetics and the Origin of Species*, there was no reason to believe that organisms were somehow "endowed with a providential ability to respond to the requirements of the environment by producing hereditary changes consonant with those requirements."[61] Evolution was not inherently progressive, and adaptation to a changing environment was anything but inevitable.

Through the work of popular writers such as Carson, the understanding of evolution articulated in the modern synthesis became linked to the environmental anxieties of the postwar period. Trained in biology at Johns Hopkins University and the Marine Biological Laboratory in Woods Hole, Massachusetts, Carson had worked for many years as a science writer for the US Fish and Wildlife Service, while also authoring a series of popular books and articles on oceanography and marine biology that by the 1950s had given her the financial independence to devote herself to writing for general audiences.[62] In the mid-1950s, she began outlining a new book on evolution before realizing that a recent book by Huxley already covered much of the same ground.[63] Around the same time, she toyed with the idea of a book on the theme of "Life and the relation of Life to the physical environment."[64] Although she abandoned these plans, the themes they addressed were central to the book she chose to write instead, *Silent Spring*. The idea that the extinction of a species occurred when its gene pool shifted too slowly to allow it to survive in a changing environment was, in Carson's view, the key to understanding why the misuse of herbicides and pesticides posed an existential threat to the human species. Humanity's success in transforming its own environment had ironically resulted in an environment that was changing at "the impetuous and heedless pace of man rather than the deliberate pace of nature," with potentially catastrophic consequences.[65]

The centrality of the modern synthesis to Carson's environmentalism is also evident in the emphasis she placed on the threat of genetic damage. Like both Huxley and Dobzhansky, she believed that humanity had a remarkable capacity to reshape its own environment and thereby affect its own evolutionary future.[66] Unlike them, however, she was pessimistic about humanity's capacity to do so wisely, particularly with the technological capacities it had recently developed. Among all the dangers to which Americans were exposed, she argued, ionizing radiation and toxic chemicals were particularly threatening due to their impact on "genetic heritage."[67] Both had the capacity to directly damage chromosomes, hinder cell division, and cause genetic mutations, bypassing many of the defenses that Carson believed humans and other organisms had evolved to protect themselves from other kinds of environmental hazards. It was for this reason that she suggested pesticides and herbicides should properly be called "biocides"—that is, not selective means of eradicating pests and weeds but rather universal killers of life.[68] By framing her argument against aminotriazole, DDT, and other "biocides" in terms of genetics, Carson gave them an extraordinary generality that they would not have had if she had focused solely on their physiological effects. Since biocides attacked genes themselves rather than just the phenotypic products of those genes, the differences in functional organization that separated humans from other species were less important than the common vulnerability of all living beings in evaluating the magnitude and nature of the threat such chemicals posed. As Carson noted in 1963 in her last public speech, delivered shortly before her death from breast cancer the following year, "Man is affected by the same environmental influences that control the lives of all the many thousands of other species to which he is related by evolutionary ties."[69] The same thing that threatened to silence the insects and birds, in other words, also threatened to silence humanity.

The modern synthesis had a similarly universalizing effect on the way environmentalists understood the significance of human racial differences.[70] In his

1962 book *Our Synthetic Environment,* for example, which addressed many of the same themes as *Silent Spring* but situated them in a more radical political framework, Murray Bookchin deployed evolutionary theory to explain why toxic chemicals and ionizing radiation posed a uniquely universal threat to humanity.[71] Whereas each human race had adapted biologically to its local climatic conditions, Bookchin argued in an echo of nineteenth-century race science, no human population had had sufficient time to evolve biological defenses against radioactive fallout, synthetic pesticides, or the industrial environment as a whole.[72] In contrast to the racially specific effects of climate, he argued, "toxicants dangerous to men of one race are equally hazardous to men of other races," with the consequence that the "appearance of these toxic agents in the atmosphere, water, and foods threatens the health of every human being."[73]

The universality of the threat to humans and other forms of life posed by the "synthetic environment" was also a result of these toxic agents' combination of ubiquity and uncertainty. While virtually everyone was exposed to some amount of synthetic toxins, it was almost impossible to predict who would suffer from that exposure and in what ways. What placed everyone in the same relation to the toxic environment was not the absolute certainty that they would all suffer the same harm but rather the shared possibility that they might suffer them at some point in their lives. Accordingly, the fable of a small town afflicted by a mysterious poison with which Carson begins *Silent Spring* is written in the subjunctive mood; it depicts not a horror that had happened but one that could happen anywhere, to anyone, if humanity were to continue along its current path.[74] Ultimately, Carson argued, the question was not whether we could prove that we had already been harmed but whether "we are being asked to take senseless and frightening risks."[75] Together with the modern synthesis, this focus on diffuse and uncertain risks helped Carson, Bookchin, and other early environmentalists articulate an argument for solidarity that they hoped would have universal appeal. If all humans were equally vulnerable to toxic pollution and other environmental threats, then all humans had a stake in reducing the risk.

Precisely which kind of response was demanded by the recognition of universal exposure to environmental risks remained open to debate, however. Whereas Carson argued for a return to nature's pace, other environmentalists saw risk-taking and technological innovation as essential to resolving the environmental and evolutionary crisis that humanity had created for itself. The bacteriologist René Dubos, for example, was just as concerned about changes to the human environment and as skeptical of narrowly biomedical approaches to health as Carson or Bookchin, but he was more optimistic about humanity's capacity to adapt, at least in biological terms, to the new environment it had created.[76] After all, he argued, a "willingness to take risks is a condition of biological success," not only for humans but for all evolving organisms, and it was clear that humanity had reaped enormous benefits from the risks it had taken in developing synthetic herbicides and pesticides.[77] On the contrary, what worried him most about environmental change was the potential for humanity's creativity to be constrained by ever-more-hostile and unforgiving surroundings—that is, an environment to which humanity might well be able to adapt biologically but in which "real human values" would be lost.[78] In the early 1960s, this position made him an uncomfortable ally of the corporate opponents of the Delaney Clause, who cited his work approvingly at the 1963 hearings inspired by *Silent Spring*.[79] Although Dubos grew increasingly worried about pollution over the course of the 1960s and '70s, he remained a "despairing optimist," convinced of the potential for humanity to adapt to its changing environment.[80]

By examining the role of the modern evolutionary synthesis in the arguments of environmentalists such as Carson, Bookchin, and Dubos, we can see how a very general kind of argument that recurs in many times and places—namely, the idea that technological advances are not always an unmitigated boon for humanity—was reformulated in terms suitable to a particular set of circumstances and aims. We can also see how such a specific reformulation could serve as the basis for a broader movement that attracted many people who had little if any knowledge of the technical details of genetics or evolutionary theory. From such books as *Silent Spring*, *Our Synthetic Environment*, and Dubos's *Man Adapting*, readers learned that modern science had proven that humanity would not automatically adapt to the world it had built for itself.[81] Whether they were pessimists or "despairing optimists" like Dubos, many were convinced that there was an important choice to be made: namely, whether the system of material abundance that had been developed in the postwar United States would benefit consumers over the long term or whether it would simply seek to satisfy their immediate desires.

FROM UNIVERSALIST ENVIRONMENTALISM TO ENVIRONMENTAL JUSTICE

While most environmentalists circa 1970 assumed that all Americans faced equal environmental risks and should therefore be equally motivated to address them, by the end of the decade some activists were challenging that assumption by pointing out that the distribution of environmental risks was highly unequal For these activists, the key issue was not pesticide- and herbicide-laden foods that threatened each consumer as much as the next, but rather toxic waste dumps that threatened only those who lived in their proximity even as they relieved others of the burdens of exposure. In the 1970s and '80s, this new generation of activists brought attention to the role of race and class in the unequal distribution of environmental risk and thereby posed a powerful challenge to the idea that there was a singular environment for a unitary human species—an idea that had often served to obscure the fact that the implicit subject of US environmentalism in the 1960s and '70s was almost always white and middle-class. Instead, these activists for "environmental justice," as it came to be known, developed new methods of research and activism to make visible how the vulnerability of certain individuals and communities to environmental risk was shaped by hierarchies of race, class, and gender.[82]

Challenges to the kind of universalist environmentalism represented by *Silent Spring* and NEPA came both from those who opposed environmental regulations of all kinds and from those who sought to ensure that such regulations were implemented in ways that were just and equitable.[83] These right-leaning and left-leaning critics of environmentalism became more vocal in the context of the "stagflation" of the mid-1970s, which combined high unemployment with a rising cost of living. Under these conditions, the Keynesian measures that had been credited with helping end the Great Depression—raising wages, investing in public works, increasing the money supply, and other ways of stimulating consumption—could not be deployed, since they also had the effect of exacerbating inflation. The result was an economic downturn that seemed, for a while, invulnerable to the usual solutions. By eroding the purchasing power of many Americans, the stagnant-but-inflationary conditions of the 1970s reawakened tensions between workers, owners, and consumers that had remained dormant as long as consumer-oriented policies could be seen as benefiting both workers and owners—the former through higher wages, the latter through higher profits.[84] In the less prosperous 1970s, environmental policies that protected consumers by imposing new burdens on producers came under attack from labor unions and business owners alike, as did the consumer protection movement more generally.[85] Reframed as a zero-sum game in a stagnating economy, environmentalism's supposedly universal appeal began to fracture.

The universalist environmentalism of the 1960s was also challenged by the growing awareness that the new environmental laws, rather than ameliorating environmental inequalities, had actually in some cases worsened them. In the case of pesticides, for example, the shift away from DDT and similar compounds clearly benefited certain animal populations and ecosystems and reduced the exposure of consumers to potentially carcinogenic substances, but the organophosphate pesticides that replaced them were in many cases even more acutely toxic to the farmers and laborers who applied them.[86] Similarly, new regulations governing the disposal of toxic waste did not protect all Americans from toxicity, as environmentalists had claimed. Instead, they often simply shifted toxicity from middle- and upper-class white communities to poorer nonwhite communities that lacked the political and economic clout to resist them. In the late 1970s and early '80s, for example, the EPA and the state of North Carolina decided to dispose of illegally dumped polychlorinated biphenyls (PCBs) in a landfill in Warren County, North Carolina, a majority-black county that was also one of the state's poorest.[87] For residents such as Luther G. Brown, the reasons the county had been selected for the dump were clear: "It's because it's a poor county—poor politically, poor in health, poor in education and because it's mostly black. Nobody thought people like us would make a fuss."[88] Local activists who protested the proposed PCB landfill found themselves in conflict not with the industries that had produced the toxic substances or the people who had disposed of them illegally—the conventional targets of environmental activism—but rather with the state and federal authorities who had decided that their community should bear the burden of cleaning them up.[89]

For activists protesting such environmental injustices, the kinds of research and expertise on which Carson and other early environmentalists had based their activism proved to be inadequate precisely because of their supposed universality. Laboratory toxicology, for example, provided estimates of risk for humans in general but said nothing about the social distribution of risks, while the official epidemiology of health departments and environmental agencies often underestimated the role of toxic risks in producing illness in marginalized communities. For reasons ranging from overly conservative scientific standards to outright corruption, environmental justice activists argued, studies conducted by state environmental agencies and the EPA were deeply compromised.[90] Consequently, activists developed their own methods, sometimes described as "popular epidemiology," to show how environmental risks became concentrated in their communities.[91] At the Love Canal housing development in New York, for example, where volatile toxic wastes from an abandoned dump had seeped into backyards and basements over years of official neglect, resident Lois Gibbs went door-to-door to collect evidence of illnesses and injuries that government agencies had proven unwilling or unable to collect.[92] On a national scale, the United Church of Christ's Commission for Racial Justice issued a report in 1987 showed that hazardous waste facilities were more likely to be sited in majority-nonwhite communities than in majority-white communities. Race, it concluded, "was consistently a more prominent factor in the location of commercial hazardous waste facilities than any other factor examined."[93] That insight served as the basis of the National People of Color Environmental Leadership Summit in Washington in 1991, an important milestone in the emergence of the environmental justice movement in the United States.[94]

The environmental justice movement also adopted organizational forms that reflected its roots in local community concerns and distinguished it from the increasingly bureaucratic mainstream environmental movement. By the 1980s, partly as a response to the new environmental laws of the late 1960s and '70s, the most influential environmental organizations had become large, centralized, and hierarchical. Headquartered in Washington, DC, they focused on filing lawsuits and advocating for changes to national policy with the help of highly trained lawyers, scientists, and other experts, turning to a broader public only when they needed donations or sought to pressure politicians through letter-writing campaigns.[95] The environmental justice movement, by contrast, evolved from scattered local efforts into a network of organizations that were united by a common set of tactics and concerns but remained centered on the histories and identities of the communities where they were rooted. Doing so preserved a space for each community to establish its own leaders, expertise, and political voice to fight environmental risks specific to that community rather than shared among all Americans, let alone all members of the human species.[96] Even when environmental justice activists established organizations with national scope, those organizations generally focused on supporting local grassroots efforts. For example, the Citizens Clearinghouse for Hazardous Waste, an organization established by Gibbs in 1981, offered tactical support and resources to communities fighting their own battles against toxics rather than trying to manage each case directly or shifting the focus to Washington. It was not a "lobby, litigation, or public interest group," it claimed, but a grassroots organization whose "main function" was "to help people help themselves."[97]

Despite the explicit embrace of the word *environment* in the name given to this emerging movement, many environmental justice activists had an ambivalent relationship to the label "environmentalist" and the concept of "environment."[98] Because existing environmental groups possessed much-needed expertise, embracing explicitly environmental language was often an effective strategy for environmental justice activists as they entered forums in which success depended on mastering the complexities of existing laws and institutions. The Love Canal Homeowners Association, for example, relied for legal advice on the Sierra Club Legal Defense Fund, which had been established by the Sierra Club in 1971 as environmentalists turned to litigation to shape the interpretation and enforcement of the environmental laws of the late 1960s and early '70s.[99] More broadly, environmental justice activists were sometimes able to attract resources, media attention, and sympathy by describing themselves as "environmentalists" even when they felt little affinity with the environmental movement as it had developed since the 1960s. At other times, however, environmental justice activists sought to distance themselves from what some of them derided as "mainstream" environmentalism.[100] This was particularly the case for activists who approached toxic dumping and other environmental issues through the frame of civil rights. People who described themselves as "environmentalists," they argued, tended to be white, affluent, and primarily concerned with the preservation of wilderness and wildlife rather than with the everyday concerns of poor, working-class, and nonwhite communities. In 1990, for example, a group of activists affiliated with the SouthWest Organizing Project wrote an open letter to leading environmental organizations accusing them of continuing to "support and promote policies which emphasize the clean-up and preservation of the environment on the backs of working people in general and people of color in particular."[101]

By examining the ways in which the concept of environment was adopted and adapted by the environmental justice movement, we can see how the emergence of the environmental movement in the 1960s and '70s changed the stakes of the term. Now associated with a specific set of issues, institutions, laws, and political allegiances, much of the interpretive flexibility of the "environment" and "environmentalism" that had characterized those terms since the mid-nineteenth century had been lost. In this context, it became harder to advocate for alternative variants of the concept of environment and the forms of environmentalism that accompanied them, since it was now necessary to point out that what one was advocating was both deeply "environmental" and clearly distinct from—indeed, sometimes diametrically opposed to—something called "mainstream environmentalism." Nonetheless, the concept of environment remained essential to environmental justice activists, who continued to argue (like the "mainstream" environmentalists who preceded them) that the consumer-centric US economy produced ubiquitous, hidden toxic risks to human health and well-being, to which no individual or population could adapt rapidly enough to avoid harm. Similarly, they continued to use the ubiquity and uncertainty of such risks as a foundation for solidarity among otherwise divergent individuals and groups. Thus, we can see how the concept of environment was materialized in a variety of ways, even as one particular way came to be associated with the term *environmentalism*.

*

The opening line of the first full chapter of *Silent Spring* fails to match some of our expectations of the environmental movement of the 1960s and '70s. It does not mention toxic poisoning or industrial malfeasance, nor does it concern itself with "nature" in conventional terms—that is, those wild places and natural resources that environmental justice activists would later accuse mainstream environmentalists of privileging to the detriment of marginalized communities. Rather, it offers the expansive claim that the "history of life on earth has been a history of interaction between living things and their surroundings."[102] Rooted in Carson's understanding of biology and evolutionary theory, this opening line lays the groundwork for the book's attack on the misuse of chemical pesticides and herbicides. What had motivated her to write the book was the recognition that the "surroundings" with which living things now had to grapple were permeated by the invisible, potentially toxic products of human ingenuity, and that the quantity and diversity of these toxic substances was increasing daily. Linked to a universalist view of humanity—that is, of humanity as a single biological species, each of whose members was equally vulnerable to toxic risks—this recognition provided the conceptual foundation for the emergence over the course of the 1960s of a social movement that was self-consciously "environmental."

By examining the role of consumer culture, toxicological research methods, and evolutionary theory in the emergence of modern environmentalism, we can see that the movement did not arise merely from the realization by astute scientists and activists of the threat posed by human activities to a preexisting object, "the environment," which they then mobilized the public and policymakers to protect.[103] Rather, it arose from the articulation of a new understanding of the concept of environment in terms of hidden toxic risks that threatened the survival of the human species, which activists used to describe environmental threats in a way that was compelling and demanded action. To do so, they relied on a range of techniques for detecting toxic substances and assessing the risks they posed, from paper and gas chromatography to human and animal testing to studies of the ecological and epidemiological effects of widespread spraying of pesticides such as aminotriazole and DDT. These "environmentalists," as they came to be known in the course of the 1960s, used this articulation of the concept of environment and the techniques that substantiated it to make their claims and to rally public support at a time when American consumers were increasingly concerned about the hidden costs of material abundance.

As the "environmental movement" grew, spurred by the massive Earth Day teach-ins of April 1970, newly self-described "environmentalists" adopted the idea of a singular humanity or community of life in relation to a singular environment.[104] Their very use of the singular term *the environment* rather than the plural *environments* reflected this universalizing turn. Defined as the conditions necessary for the survival and well-being of the human species or of life as a whole, "the environment" offered a powerful new discourse for establishing solidarity across borders of class, race, gender, and nationality—a universalist environmentalism that could and did claim to be relevant to everyone. This universalism was also a source of weakness, however, inasmuch as it led many environmentalists to ignore differences in the environmental exposures of different individuals and groups, as well as the varied goals and values that different individuals and groups brought to their relationships to their surroundings. These blind spots led in some cases to direct harm, as in the case of Warren County's EPA-supported toxic waste dump, and in other cases to skepticism about the aims of environmental activists. Universalist claims about improving "the environment," environmental justice activists made clear, had often been used to advocate actions that improved only the environments of particular groups.

Paying attention to how the concept of environment has changed over time and from one community of speakers to another also helps us gain a clearer view of the environmental movement of today. In particular, it reveals how the idea of a universal humanity confronted by a singular environment has been at the very same time one of the movement's greatest strengths and one of its greatest weakness. The environmentalism of the 1960s proved compelling across the political spectrum because it posited the existence of risks that affected all humans regardless of class, race, gender, or political orientation. To do so, it offered not just a new evaluation of the state of the environment, but also a new understanding of the concept of environment. In the United States, this vision of environmental quality proved easier to sustain from the late 1950s to the early 1970s, when postwar prosperity was at its height and economic growth could plausibly be argued to be benefiting all Americans, than it did from the mid-1970s onward, when a stagnating economy and rising inequality revealed the deep divides between rich and poor, black and white, producer and consumer. The next chapter turns to the closing decades of the twentieth century, when a different set of tensions—between local and global views and between the Global North and Global South—shaped the adaptation of the concept of environment to a world being reshaped by climate change.

CHAPTER 6

The Human Planet:
Globalization, Climate Change, and
the Future of Civilization on Earth

Carbon dioxide is a colorless, odorless gas capable of absorbing some of the heat radiated from Earth's surface after it has been warmed by the sun. This makes it one of the so-called greenhouse gases (along with methane, water vapor, and others) that keep the planet's atmosphere, oceans, and land at temperatures hospitable to life. Over the history of Earth, the levels of these gases have varied, as has the planet's temperature. For the vast majority of that history—indeed, for the vast majority of the very thin slice of that history during which humanity has existed—these fluctuations have had nothing to do with human activities. In 1990, however, the Intergovernmental Panel on Climate Change (IPCC) released a report warning that the human use of fossil fuels such as coal, oil, and gas had increased the atmospheric concentration of carbon dioxide to levels that would probably soon lead to a significant increase in global temperatures, if they had not already.[1] Over the previous century, the IPCC found, global temperatures had already increased by 0.3 to 0.6 degrees Celsius, while the average carbon dioxide concentration had risen from about 290 parts per million (ppm) to more than 350 ppm.[2] The report concluded that it was still too early to say whether the increase in global temperatures was caused by, rather than merely correlated with, the increase in carbon dioxide and other greenhouse gases. Nonetheless, given the potentially catastrophic consequences of global warming, it stressed the urgent need for an international research program "to improve our capability to observe, model and understand the global climate system."[3]

Concerns about the impact of fossil fuels on the global climate had been growing among climate scientists for decades, but the IPCC's consensus-building approach gave its report an authority that previous warnings had lacked. Following its release, a wide range of environmentalists and policy-makers pushed not only for more research but also for immediate action to minimize the risks of rapid climate change. In 1992, after several years of ne-gotiations, the United Nations Framework Convention on Climate Change (UNFCCC) was opened for signatures at the UN Conference on Environment and Development in Rio de Janeiro.[4] The treaty went into force two years later, after it had been ratified by the first 50 of what would eventually grow to 197 countries and other parties, including all member states of the United Nations. Although parties to the convention did not commit to reducing their emissions or taking other immediate steps, they agreed that "change in the Earth's climate and its adverse effects are a common concern of humankind" and that human greenhouse gas emissions would "result on average in an ad-ditional warming of the Earth's surface and atmosphere and may adversely affect natural ecosystems and humankind."[5] They also affirmed their intent to "protect the climate system for the benefit of present and future generations of humankind, on the basis of equity and in accordance with their common but differentiated responsibilities and respective capabilities."[6]

The transformation of climate change from the obscure research topic of a few specialists to one of the most important subjects of international diplo-macy and environmental activism in the last quarter of the twentieth century had multiple causes. One was a dramatically improved understanding of the global climate system. Since the early decades of the Cold War, climate scien-tists had expanded the reach of their data-gathering networks, refined their computational models, and created new institutions to facilitate the sharing and evaluation of data and models. They also developed a new understanding of the global climate as an object of knowledge and concern. Whereas most earlier climatologists had understood "climate" as a property of particular re-gions of Earth that could be studied through direct observation and measure-ment, late-twentieth-century climate scientists focused on the "climate sys-tem" as a singular, planetary phenomenon that could be studied only through globe-spanning sensor networks and computer simulations.[7] By the time of the IPCC's founding in 1988, a few climate scientists had concluded that this global climate system was already being transformed by human activities. While initially a minority position, over the succeeding decades this hypothe-sis was embraced by a growing number of scientists. By the end of the century, virtually all researchers active in the field agreed that the climate was already changing and that humanity's use of fossil fuels was the primary cause.[8]

Scientific advances were not the only reason for climate change's rise to prominence during this period, however. Equally important were the era's geopolitical shifts—particularly, the fall of the Berlin Wall in 1989 and the collapse of the Soviet Union in 1991, which brought the nearly half-century-long Cold War to an end. In its wake, the ideological clash between US-style democratic capitalism and Soviet-style authoritarian socialism that had dominated international relations since the late 1940s was replaced by new tensions between the developed nations of the Global North and the developing nations of the Global South. With some form of capitalism adopted as the economic system virtually everywhere, those tensions centered not on attempts to revolutionize the existing system but on structural factors that led some to prosper within it while others remained desperately poor. In this post–Cold War context, climate change and the various efforts to mitigate it became an important arena for contests over the future of the system. While the threat of global warming made it clear that the world's nations were bound together by the planet they shared, it also made it clear that they were divided by vast differences in responsibility, vulnerability, and capacity to respond. Proposed solutions to the risks posed by climate change were therefore often evaluated both in terms of their ability to slow the planet's warming and in terms of how they would restructure the post–Cold War global order.

Climate scientists and activists refashioned the concept of environment to suit these new needs and circumstances. Before global warming became a matter of widespread concern in the 1980s, few had considered it to be an "environmental" issue, at least in the sense that the term was commonly used among environmentalists of the time, who focused on threats such as toxic pollution that were potentially harmful to individuals and communities everywhere. Climate change was simply too uncertain, too large in scale, too delayed in time, and too variable in its local effects to serve as a compelling basis for an environmentalism of this kind. In the 1970s and '80s, however, climate scientists and activists began to articulate a new way of understanding global climate change in explicitly environmental terms. Rather than focusing on its consequences for individuals, communities, ecosystems, or even the human species as a whole, they focused on the role played by the global climate in sustaining something they called "human civilization" or "the human enterprise." As the planet warmed, they argued, the expansion in the scope of human freedoms and comforts that they associated with the development of modern civilizations would become increasingly constrained, and indeed certain of those freedoms and comforts might be lost. At worst, civilization itself might collapse. While this vision of an existential threat to human civilization gave activists a common focus, it also masked many of the tensions and divisions of the era. In the early decades of the twenty-first century, as attempts to reduce greenhouse gas emissions through international regulations faltered, these tensions led some people to propose alternative, more pluralistic ways of envisioning the global environment.

GLOBALIZING

When the IPCC concluded in its First Assessment Report in 1990 that the burning of fossil fuels might already be changing the global climate, its ability to make a convincing case was the result of two developments in the second half of the twentieth century: a dramatic rise in the emission of greenhouse gases from human activities, and dramatic improvements in scientists' ability to measure those emissions and model their effects. Undergirding both of these developments was something that might be called "globalization," as long as that term is understood broadly to mean the increasing entanglement of the lives and livelihoods of people living and working at great distances from one another. This entanglement took many forms in the decades after World War II. It was military, in that the two nuclear-armed superpowers of the Cold War era—the United States and the Soviet Union—approached the entire planet as a potential battlefield. It was also economic, in that both of those powers sought to integrate the nations within their respective blocs into global systems of production and consumption. It was technological, cultural, and political, in that infrastructures of transportation and communications were used to spread ideas and values globally and to knit together new political alliances that transcended national borders. Finally, globalization during this period was scientific, in that scientists were able to use new kinds of global institutions and infrastructures to better understand the planet as a whole.

For climate scientists in particular, globalization in the post–World War II period provided the foundation for the development of new data-gathering networks, computational models, and research collaborations that helped them grasp the global climate system as an object of knowledge and concern. This was not, of course, the first time that scientists had built international networks to gather data on a scale that was beyond the reach of each of them alone. In the nineteenth century, scientists such as William Whewell had attempted to map the world's tides and climates using data collected by hundreds or even thousands of observers. By the end of the nineteenth century, moreover, cooperation among the scientists of different nations had already been formalized through institutions such as the International Meteorological Association (IMA), which later served as the basis for the World Meteorological

Organization (WMO), founded under the auspices of the United Nations in 1951.[9] In the case of meteorology and climate science, what changed between the late nineteenth century and the mid-twentieth was not the aim of coordinating large-scale, geographically dispersed research but rather the way such research was coordinated. The IMA's "international" association of national weather bureaus, each of which shared methods and ideas with the others in order ultimately to better understand its own problems, gave way to the "world" organization of the WMO, which sought to manage the collection and distribution of weather and climate data on a global scale.[10]

Among the drivers of this shift from international to global science were the military and geopolitical interests of the two Cold War superpowers, which helped create the professional and technological infrastructure for the global environmental sciences in the second half of the twentieth century.[11] Following World War II, both the United States and the Soviet Union invested heavily in scientific research that would help them monitor each other's nuclear weapons programs, plan for warfare on land, in the air, underwater, and in space, and manipulate the environment to help allies or harm enemies (by, for instance, controlling the weather or disseminating crop diseases). As a result, the amount of funding available and the number of researchers at work in fields such as seismology, oceanography, meteorology, geophysics, and ecology expanded dramatically. Although some of this research was explicitly military in nature, much of it took place in the context of civilian projects such as the International Geophysical Year of 1957/58 or at institutions such as the Scripps Institution of Oceanography in La Jolla, California, which was nominally civilian but funded almost entirely by the US Navy.[12] From a national security perspective, what mattered was not so much the acquisition of a particular fact or the development of a specific scientific theory as the availability of specialized expertise that could be applied to military needs as necessary. Much of the early work on modeling of the climate system was of this nature. While the immediate military relevance of research on basic processes of atmospheric and oceanic circulation was often unclear, the value of having a group of experts on these likely "environments" for future war seemed obvious to military planners.[13]

Beyond such national security concerns, the development of new technologies of transportation, communications, and computing for both military and civilian purposes also provided tools that scientists could use to study climate on a global scale. The expansion of international air travel, for example, made it easier for scientists to carry out measurements or install sensors and other equipment at distant sites, while telecommunications technologies—from transcontinental telephone lines to satellite links to the internet—facilitated the rapid

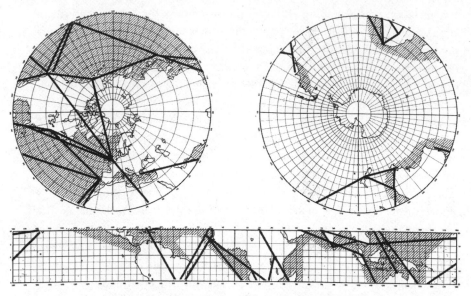

FIGURE 12. A global map of areas where the scientists involved in the First GARP Global Experiment (1978/79) planned to gather data from merchant ships (shaded) and commercial aircraft (bold lines). (Reprinted from figure 6.2 in Joint Organizing Committee of the Global Atmosphere Research Programme, *The First GARP Global Experiment: Objectives and Plans*, GARP Publications Series, no. 11 [Geneva: World Meteorological Organization and International Council of Scientific Unions, 1973], 17, with permission from the World Meteorological Organization.)

transfer of data from field sites to central repositories as well as the sharing of data and models between scientists. The launching of weather and communications satellites in particular gave climate scientists new tools both for measuring Earth's atmosphere from above and for collecting data in near-real-time from ground-, sea-, and air-based sensors. The Global Atmospheric Research Programme (GARP), launched by the WMO in 1967 to standardize and coordinate data-gathering on weather and climate, for example, included several major "experiments" that collected data on a global scale from weather stations, buoys, ships, balloons, and satellites.[14] For the First GARP Global Experiment, for example, scientists relied heavily on the new weather satellites and dedicated weather stations as well as on measurements of wind speed, temperature, and other parameters acquired from commercial aircraft and merchant ships.[15] They also took advantage of advances in computing technology—which were first driven mainly by military concerns and then by the development of a commercial market—to develop ever-more-complex models of the global climate system.[16]

As scientists globalized their studies of weather and climate, they also played a critical role in the emergence of a new set of institutions that sought both to accelerate globalization and to monitor and manage its environmental impacts. Most of the international institutions founded in the immediate postwar period (including the UN Food and Agriculture Organization and the World Bank) focused on promoting forms of economic development that would link the developing world more tightly into the global trade system. These development organizations were joined over succeeding decades by others that focused on the problems and inequities of development, including its negative environmental impacts. An important turning point was the 1972 UN Conference on the Human Environment in Stockholm, which led to the establishment of the UN Environment Programme (UNEP) in Nairobi and firmly established the environment as a subject of international diplomacy.[17] Although UNEP was as explicitly pro-development as other UN agencies, its support was tempered by a concern for the environmental causes and consequences of inequality among nations. This tempered developmentalism was reflected in one of its founding aims, which was to "effectively assist developing countries to implement environmental policies and programmes that are compatible with their development plans."[18] The same attempt to balance development and its environmental impacts shaped the 1987 report by the World Commission on Environment and Development, *Our Common Future*, now remembered best for its introduction of the term *sustainable development* into widespread use.[19] That report, in turn, influenced the 1992 UN Conference on Environment and Development, where representatives of the world's nations agreed to "protect the integrity of the global environmental and developmental system," including the climate system described in the IPCC's First Assessment Report.[20]

Globalization in the second half of the twentieth century also had material effects on the world that scientists were observing. The human impact on the planet accelerated in a variety of ways, including the appropriation of an ever-greater proportion of the land surface and fresh water for agricultural purposes, the extinction of numerous species, and the transformation of key biogeochemical cycles, including the carbon cycle. The cause of this "Great Acceleration," as it has sometimes been called, was not simply that the human population had risen to more than six billion by the end of the twentieth century or that an increasing proportion of that population was consuming energy and goods at unsustainable rates even as global inequality deepened.[21] It was also that the connections between people living and working at great distances from one another grew so dense, rapid, and diverse that it became difficult if not impossible to understand local changes in the material conditions of life without situating them in relation to faraway places and global processes. For example, the transformation of the ecosystems of the Great Lakes of North America by invasive mollusk and fish species was the result of global shifts in the economics and technology of shipping, while rising seas caused by the greenhouse gas emissions of developed nations threatened to displace coastal villages in Alaska and the Maldives alike. These were the results not just of growing populations and levels of consumption but also of thickening webs of connection among distant places and peoples.

The climate science of the late twentieth century would not have developed without the technological, economic, cultural, and political forms of globalization that were also responsible, during the very same period, for dramatically expanding humanity's environmental footprint. It therefore offers a particularly compelling example of a more general truism—namely, that improvements in our understanding of the world around us depend on our relationship to that world, and that relationship is always changing in ways large and small. It might be possible to imagine a scientist capable of observing the world without becoming entangled with it; in practice, laboratories must be built, personnel hired, instruments purchased, experiments conducted, field sites visited, and samples collected. None of these activities can be successfully pursued in isolation. Moreover, when the object of research is a phenomenon that is global in scale, the practical work required to study it is, in some form or another, necessarily global as well. It is therefore hardly a surprise that the global connections that made the Great Acceleration possible were also essential to the production of knowledge about the effects of that acceleration on the planet.

MODELING

Although the methods, data, and theories that constituted climate science in the late twentieth century were novel in many ways, the possibility that humanity could change the climate on both local and global scales had already been discussed by scientists for centuries, if not longer. As early as 1778, Buffon had suggested in his book *The Epochs of Nature* that by planting trees, setting fires, redirecting rivers, and otherwise altering the landscape, "man can modify the influences of the climate that he inhabits, and set it to the temperature that suits him."[22] Over the following century and a half, a number of scientists suggested that, in particular, carbon dioxide emitted by human use of fossil fuels might be capable of absorbing heat from the sun and thereby raising the planet's temperature.[23] Nonetheless, few scientists in the mid-twentieth century were alarmed. For one thing, there was no reliable evidence of an increase in either of the two critical variables, global temperature or atmospheric carbon dioxide concentration. For another, it was obvious to all that scientists' models of the climate system were highly oversimplified, omitting feedback processes that seemed fully capable of neutralizing any warming effect.[24]

Moreover, as scientists refined their understanding of the relevant physics, the theoretical difficulties with the idea of fossil fuel–driven global warming seemed to grow rather than shrink. Before the mid-twentieth century, most scientists were convinced that a majority if not all of the carbon dioxide generated by human activities would be absorbed by the oceans, and moreover that the spectrum of thermal radiation blocked by carbon dioxide overlapped with that of water vapor, which was far more abundant in the atmosphere and would therefore render moot the effects of any increase in carbon dioxide levels. Finally, even among scientists who believed that rising carbon dioxide could lead to global warming, none were especially concerned. On the contrary, they saw it as an unintended but welcome byproduct of using fossil fuels to generate heat and power for human purposes. The British engineer and self-taught climate scientist Guy Stewart Callendar, for example, argued in a 1938 paper that increasing carbon dioxide concentrations would boost crop yields globally, while rising temperatures would expand agriculture in northern regions and delay what he described as the "return of the deadly glaciers."[25] In short, most mid-twentieth-century scientists did not believe that humanity was physically capable of warming the planet, and those who did thought there was little to fear from rising temperatures.

Only in the 1950s, when Earth scientists revised their understanding of carbon dioxide as a greenhouse gas in ways that made it clear that burning fossil fuels could cause both carbon dioxide levels and global temperatures to rise, did this mixture of skepticism and complacency begin to shift. Many of these findings emerged from institutions funded by the US military, including the Scripps Institution of Oceanography, then headed by Roger Revelle. In a landmark 1957 paper, Revelle and the physical chemist Hans Suess argued that the oceans were incapable of absorbing carbon dioxide quickly enough to prevent human activities from raising its concentration in the atmosphere.[26] Indeed, they calculated, the average atmospheric carbon dioxide concentration was already 3 to 6 percent higher than it would have been without the burning of fossil fuels. Drawing on UN projections of future fossil fuel use, they predicted that the carbon dioxide concentration would rise 20 to 40 percent in the coming decades—an amount that would be "adequate to allow a determination of the effects, if any, of changes in atmospheric carbon dioxide on weather and climate throughout the earth."[27] Around the same time, the physicist Gilbert Plass published a series of papers showing that an incomplete understanding of carbon dioxide's absorption spectrum and the way it was distributed throughout the air column had led scientists to dramatically underestimate its effectiveness as a greenhouse gas.[28] The result was that by the late 1950s the two main theoretical obstacles to accepting the geophysical possibility of anthropogenic climate change had been eliminated, making it easier for scientists to believe both that humanity could cause carbon dioxide levels to rise and that those rising levels could trap heat in the atmosphere that otherwise would have radiated into space.

Nonetheless, determining precisely how much global warming could be expected from a given increase in carbon dioxide concentrations and how that warming would affect temperature and weather in specific places required much more sophisticated models of the global climate than existed at the time. To develop such models, scientists drew on an area of research known as numerical weather prediction (NWP) that had expanded in the years following World War II as meteorologists adapted computers originally built to simulate nuclear explosions to the purpose of forecasting the weather.[29] Such simulations tracked air currents and energy transfers from one part of the atmosphere to another in quantitative terms. Though originally designed to model short-term perturbations in local and regional weather patterns, with suitable modifications these techniques could also be used to model the long-term evolution of the global climate system. The first such general circulation model (GCM) of the global climate was developed in the mid-1950s by the meteorologist Norman Phillips at the Institute for Advanced Study in Princeton, one of the main centers for NWP research.[30] Though highly simplified, Phillips's model was capable of reproducing the broad features of the global climate system using just a few basic physical equations. Its success inspired a number of more ambitious efforts by other research groups over the following decades.[31] While initially crude, these GCMs became more powerful as meteorological datasets became more densely and evenly distributed across the globe, as scientists refined their understanding of the physical processes involved, and as faster computers made it possible to enhance the spatial and temporal precision of the models without slowing them beyond the point of usability.[32]

To test and refine these increasingly complex models, climate scientists sought out new sources of data that could reveal relationships between greenhouse gases and climate both in the present day and over the course of Earth's geological history. For the most recent periods, they could use direct measurements of weather and temperature acquired by a growing global network of standardized weather stations as well as from sensors attached to buoys and satellites. The WMO's GARP initiative, which ran from the late 1960s to the early 1980s, was one important source of such data. Climate scientists also took advantage of detailed records of changing atmospheric chemistry, such as the measurements of carbon dioxide levels at the Mauna Loa Observatory in Hawaii begun by Charles David Keeling during the International Geophysical Year of 1957/58, when annually averaged carbon dioxide levels were around 315 ppm.[33] Continued over decades, Keeling's measurements revealed a steady increase in average annual carbon dioxide levels at a rate of about 1.5 ppm per year.[34] To reconstruct the relationship between atmospheric chemistry and climate for periods that were hundreds, thousands, or even millions of years in the past, scientists developed other methods, including the analysis of samples taken from ice cores, tree rings, and other proxies, some of which could be used to reconstruct not only past carbon dioxide levels but also the temperatures and weather patterns with which they were correlated.[35]

For the sake of simplicity, climate scientists initially used GCMs to model the atmosphere alone, ignoring the interactions with the oceans and sea ice that they knew were crucial to the global climate system. By the 1970s, however, they were able to take advantage of faster computers, new data, and better understandings of the basic physical processes to develop so-called "coupled" models of oceans, atmosphere, and ice that were capable of predicting the global and regional consequences of increasing carbon dioxide levels. In 1975, for instance, a group led by Syukuro Manabe at Princeton's Geophysical Fluid Dynamics Laboratory developed the first such coupled model to include realistic representations of the shapes of oceans and continents.[36] At the same time, climate scientists were developing GCMs that accounted for the complex role of atmospheric phenomena such as cloud cover and aerosol pollution, which were capable of both potentiating and mitigating carbon dioxide's warming effect on the global climate system. At the National Center for Atmospheric Research in Boulder, Colorado, for example, Stephen Schneider developed a climate model that incorporated the cooling effect of airborne particles that reflected some portion of the sunlight striking the atmosphere before it could reach the surface. While Schneider's model initially predicted that the cooling effect of aerosol pollution produced by burning fossil fuels might outweigh the warming effect of carbon dioxide, by the late 1970s it was clear that warming would dominate.[37]

However complex such coupled models of oceans and atmosphere might have become, they continued to leave out another set of important feedback mechanisms—namely, those arising from the responses of living beings to the changing atmosphere and climate. In principle, for example, climate scientists knew that rising carbon dioxide could lead to faster vegetation growth, which would simultaneously absorb excess carbon dioxide in the atmosphere and reduce the reflectivity of the land surface. Both of these effects were capable of changing the relationship between fossil fuel use and global warming, although in different directions—the reduction in carbon dioxide would decrease amount of heat trapped on the planet, while the decrease in reflectivity would increase it. Determining the precise strength of this and other interactions between ecosystems and the climate system was essential to predicting climate change, but it was also well beyond the scope of existing models. In the 1980s, closing that gap became one of the main aims of the new interdisciplinary field of Earth system science (ESS).[38] Using techniques of remote sensing, computer modeling, and planetary science, ESS integrated studies and methods from across the physical and biological environmental sciences to understand the Earth system as a whole. Heavily promoted by the National Aeronautics and Space Administration in its early years, ESS gained a wider and more international base of support in 1987 with the establishment of the International Geosphere-Biosphere Programme (IGBP), a cooperative scientific endeavor that eventually enrolled thousands of scientists from around the world. In the words of its first executive director, the Swedish soil scientist Thomas Rosswall, the IGBP aimed "to describe and understand the interactive physical, chemical, and biological processes that regulate the total Earth system, the unique environment that it provides for life, the changes that are occurring in this system, and the manner in which they are influenced by human action."[39] Eventually continued for nearly three decades, the program helped coordinate a global network of scientists interested in the global climate as part of the broader study of the Earth system.

Even within the broad ambit of the IGBP, one important aspect of the future of the Earth system remained difficult to model in a rigorous way: the future actions of humanity itself. In the early years of global climate modeling, scientists estimated future inputs of carbon dioxide and other greenhouse gases into the climate system on an ad hoc basis using projections of population growth and economic development from institutions such as the UN Population Division and the World Bank. Since climate scientists were interested in knowing how changes to "business as usual" might affect climate, however, they also had to take into account the possible effects of proposed changes in laws, technology, economics, and land-use practices. In the late 1980s and early '90s, the IPCC formalized these projections in the form of official "scenarios," each of which represented a different hypothesis about how humanity as a whole would respond to the possibility of dramatic climate change over the coming century, ranging from business as usual to drastic cuts in emissions.[40] These scenarios were, as the IPCC admitted, "inherently controversial because they reflect different views of the future."[41] Over the course of the 1990s, economists such as William Nordhaus developed quantitative models of the feedback between climate change and economic activity in an effort to make such projections more rigorous and less controversial.[42] Known as integrated assessment models (IAMs), these models predicted how climate change would affect greenhouse gas–emitting economic activities on regional and global scales and how those economic shifts would in turn affect the pace of climate change.

By following the progressive expansion and increasing complexity of global climate models from their beginnings in Phillips's proof-of-concept GCM in the late 1950s to Nordhaus's economically informed IAMs half a century later, we can see how models of the climate system were gradually transformed into simulations of the relationship between human civilization and the conditions it required to survive. At the beginning of this period, even though scientists had long suspected that human actions could influence the climate and vice versa, they did not attempt to incorporate into the models themselves any representation of the relationship between the organization of human societies and the fate of the Earth system. In those early models, human activities were reduced to external "inputs"—that is, estimates of the contribution of fossil fuel use to atmospheric carbon dioxide levels—with no attempt to model the reciprocal effects of climate on human societies. By the end of this period, at least some models, however reductive and contested they might still have been, were attempting to simulate the dynamic, reciprocal, mutually constitutive relationship between human activities and the Earth system. In other words, scientists had begun to model the climate not just as an aspect of the physical Earth system but as an "environment" for human life.

MITIGATING

Even with the advances in modeling the climate system and its interactions with the biosphere and human activities that took place between the 1950s and the 1980s, most climate scientists still hesitated to make definitive claims about the magnitude or the timing of anthropogenic global warming, aware as they were of the complexity of the climate system and the continuing limitations of their data and models. Nonetheless, by the early 1970s, they knew enough to worry that the burning of fossil fuels might sooner or later have disastrous consequences for the climate and for humanity, and some of them began to search for forums where they could translate their concerns into action. In preparation for the 1972 UN Conference on the Human Environment in Stockholm, for example, climate scientists contributed to several reports on the "global environment" that singled out the climate for special concern. One of these reports, prepared by the Scientific Committee on Problems of the Environment (SCOPE), emphasized the importance of establishing a "global environmental monitoring" system capable of improving humanity's understanding of global warming and other threats.[43] At the Stockholm meeting itself, imagery of the "whole Earth" and "one world" was ubiquitous, and climate change was discussed alongside other global environmental problems such as air pollution, water scarcity, overpopulation, and extinction. After the meeting, a quasi-official report by Barbara Ward and René Dubos, *Only One Earth*, included a section on emerging threats to the climate, oceans, and other aspects of the global "planetary order."[44] As these developments suggest, global warming had clearly found a place on the international policy agenda as an environmental threat to humanity as a whole by the early 1970s.

That place was nonetheless an uncertain one, particularly as the enthusiasm around the Stockholm meeting faded and attendees returned to the domestic political contexts in which they conducted the majority of their work. As Ward and Dubos noted in *Only One Earth*, environmental problems that were fundamentally global in nature—that is, that were inherently the product of global-scale processes rather than simply being local-scale problems that happened to be visible everywhere on the globe—were difficult to address both through the domestic policies of individual nations and through conventional international diplomacy. Humanity's "global interdependence begins to require, in these fields, a new capacity for global decision-making and global care," they argued, which in turn depended on the emergence of a sense of "planetary community and commitment."[45] Ward and Dubos expressed hope that such a sense of planetary community was already emerging, and they pointed to the Stockholm meeting as an important milestone in its development. Nonetheless, the practical challenges faced by various national environmental movements and the tensions among geopolitical and economic blocs left little space for global concerns.[46] With only a few exceptions, even the most globally oriented environmentalists concentrated instead on problems such as pollution and overpopulation that either affected all humans equally or that had identifiable local causes and consequences. As a global problem with varied and unpredictable consequences for individuals and regions— that is, a problem that was neither universal nor local in the senses important to the environmental movement—climate change was an awkward fit to the environmentalism of the era.[47]

Following the Stockholm meeting, even those climate scientists who were willing to take a public stance on the urgency of climate change hesitated to call for immediate action, instead focusing their efforts on improving their basic understanding of the climate system in the hope that refinements to the science would eventually make clear the magnitude of the threat. Over the course of the 1970s and '80s, this "science first" approach led to growing confidence among scientists that climate change needed to be taken seriously and that the best way to do so was to bring experts into direct contact with policymakers.[48] That was the main aim of the IPCC, which was established in 1988 as a joint project of UNEP and the WMO.[49] Aware that climate science was extraordinarily complex and that any recommendation to curtail fossil fuel use globally would be highly controversial, the IPCC sought to secure the scientific and political legitimacy of its conclusions by bringing scientists together with representatives of national governments to craft its assessment reports, each of which was the result of a multiyear process of negotiation and revision. This painstaking process was designed, in the words of the IPCC's first chairman, the Swedish climate scientist Bert Bolin, to maintain the scientific integrity of the assessment while also coordinating "the interactions between the scientific community, stakeholders and politicians that might bring the issue forward politically."[50] In principle, if not always in practice, its procedures ensured that all nations and political positions were heard and that none would be allowed to exercise an undue influence.

The rapid progress from the IPCC's First Assessment Report in 1990— which was widely accepted as an authoritative statement of the state of the art in climate science—to the drafting and ratification of the UNFCCC a few years later seemed to confirm the wisdom of this approach, but it also proved to be its last unqualified success. The limits of an internationalist, science-first approach became apparent at the 1997 Conference of the Parties to the UNFCCC in Kyoto, where negotiators sought for the first time to establish binding international regulations on greenhouse gas emissions. The Kyoto Protocol adopted at the end of the conference called for developed nations to reduce their average annual emissions of greenhouse gases to 5 percent below 1990 levels by a "commitment period" that ran from 2008 to 2012.[51] Despite the fact that the establishment of a binding target of any kind was a major accomplishment, the protocol fell far short of climate activists' hopes. Even if its targets were met, they recognized, it would do little to prevent global warming, since reducing emissions to 5 percent below 1990 levels would still leave them high enough to continue raising atmospheric concentrations of greenhouse gases. Moreover, the treaty placed no restrictions on the emissions of developing nations such as China, India, or Brazil, which had vehemently opposed any attempt to restrict their economic growth in order to solve a problem they believed had been caused by the prior industrial development of Europe, North America, and Japan.[52] Finally, at the urging of the United States—which opposed any provisions that challenged free market principles or threatened its global dominance—the protocol also included a number of "flexibility mechanisms" that allowed developed nations to pursue carbon-intensive "business as usual" while transferring wealth to developing nations that followed low-carbon development pathways. These included the Clean Development Mechanism (CDM), under which developed nations could fulfill their treaty obligations by paying developing nations to reduce carbon emissions or increase carbon "sinks" that removed carbon dioxide from the atmosphere (for instance, by planting trees).[53]

Over the years following the Kyoto conference, it became clear that the protocol's modest targets, extensive exclusions, and liberal flexibility mechanisms had rendered it incapable of achieving its putative aim of slowing the pace of climate change. Even before it entered into force in 2005, having been ratified by a sufficient number of parties (but not the United States), its target reductions in carbon dioxide emissions were widely recognized as inadequate, as were its enforcement mechanisms. By the commitment period of 2008–2012, the developed nations subject to the protocol had in fact managed to meet their targets, but some of them met it only with the help of the protocol's flexibility mechanisms, and all benefited from the global economic slowdown resulting from the financial crisis of 2008.[54] In other words, they met their targets mainly by paying for trees to be planted in developing countries and temporarily shuttering factories at home, not by permanently shifting to a low-carbon economy. Meanwhile, China—a developing nation that was not required to limit its emissions under the treaty—was estimated to have overtaken the United States as the world's biggest greenhouse gas emitter sometime around 2006.[55] Thus, the protocol had virtually no effect on the two nations that were most critical to mitigating climate change: the United States because it refused to ratify the treaty despite having successfully negotiated to weaken it, and China because it was explicitly excluded from the treaty's most stringent requirements. As scientists, activists, and diplomats prepared for a major Conference of the Parties to the UNFCCC scheduled for Copenhagen in 2009, it was evident that the two global institutions created to establish consensus on climate change and to negotiate a global response to it, the IPCC and the UNFCCC, were foundering on the unresolved geopolitical tensions of the post–Cold War world.

As hope faded that the consensus-building, science-first, top-down approach of the IPCC and the UNFCCC would suffice to mobilize a global response in a divided world, a new wave of climate activist organizations emerged to promote a different strategy. Although these organizations continued to support the kind of binding global agreements they hoped would be negotiated in Copenhagen, they put less faith in the power of climate science by itself to compel a reasonable and effective response, no matter how credible that science might be. Instead, they sought to mobilize a global grassroots movement. Among the most successful of these new organizations was 350.org, founded in 2008 by the writer and activist Bill McKibben and a group of former students at Middlebury College in Vermont after a decade of growing frustration with the failures of post-Kyoto climate policy.[56] The origins of 350.org lay in a "climate march" to Vermont's state capital that McKibben organized in 2006 to demonstrate that the climate was something that concerned not just scientists and diplomats but also ordinary citizens. Following the success of the march and similar actions elsewhere in the United States, McKibben and his cofounders decided to launch a new organization that would harness the power of the internet to link dispersed demonstrations around the world with a unified call for climate action. They chose a name, 350.org, that they believed would translate easily across linguistic and cultural barriers while also invoking the results of a recent study by the climate scientist James Hansen and his colleagues, who had argued that humanity risked "irreversible catastrophic effects" if it failed to reduce atmospheric carbon dioxide levels below 350 ppm.[57] 350.org's first major campaign was an International Day of Climate Action on October 24, 2009, through which it hoped to influence negotiators at the upcoming UNFCCC conference in Copenhagen. Epitomizing 350.org's broader strategy, the campaign used photographs, images, and messages shared over the internet to link thousands of demonstrations—ranging in size from a few people to several thousand—into a single virtual protest on a global scale.[58]

By tracing the development of climate activism from the Stockholm conference in 1972 to the Copenhagen Conference of the Parties to the UNFCCC in 2009, we can see how scientists, activists, and policymakers attempted to materialize the "planetary community and commitment" that Ward and Dubos had argued was essential to solving global environmental problems. There were some notable successes, including the creation of the IPCC in 1988 and

FIGURE 13. Protestors at a demonstration in Toronto organized to coincide with 350.org's International Day of Climate Action on October 24, 2009. (Photograph by Flickr contributor Tania Lui, accessed May 6, 2019, https://www.flickr.com/photos/75511860@N00/4041007007/, published under Creative Commons Attribution-NoDerivs 2.0 Generic License [CC BY-ND 2.0].)

the adoption of the UNFCCC in 1992. Largely, though, this was a story of failure. In the geopolitical context of the late Cold War and its immediate aftermath, any sense of planetary community was trumped by tensions between the Global North and the Global South. It was clear that the relatively wealthy residents of developed countries both benefited from and contributed to their own nations' outsize role in causing global warming, for example, while often being insulated from its worst effects through their geographical good fortune and through their wealth. It was just as clear that the relatively poor residents of developing countries—who had neither contributed significantly to global warming nor benefited much from fossil fuel–based industrialization—were usually the most vulnerable. Even the relative success of nongovernmental organizations such as 350.org in mobilizing global protests failed to overcome such entrenched interests and deep-seated differences. By the early twenty-first century, global humanity had therefore become an important object of knowledge and concern among scientists and activists, but it had failed to materialize as a political actor capable of taking responsibility for its own conditions of survival.

THE GLOBAL ENVIRONMENT OF THE
HUMAN ENTERPRISE

In the early decades of the twenty-first century, as climate scientists and activists struggled to articulate an effective response to the globally connected but locally variable threat of climate change, they adopted a concept that had first been used in print just a few years before: the "Anthropocene." The first appearance of the term in print came in 2000, when the Dutch atmospheric chemist Paul Crutzen and the US aquatic biologist Eugene Stoermer published a brief article on the idea in an issue of the IGBP's *Global Change Newsletter*.[59] At some point in the previous several centuries, Crutzen and Stoermer argued, the Earth system had left behind the Holocene epoch, which geologists understood to have begun about 11,700 years ago with the end of the last ice age. In its place was a new geological epoch, the Anthropocene, which was defined instead by what they described as "major and still growing impacts of human activities on earth and atmosphere, and at all, including global, scales."[60] As a tentative starting point for the Anthropocene, Crutzen and Stoermer chose the late eighteenth century, when humanity's global impact was just becoming visible through increases in atmospheric concentrations of carbon dioxide and methane and when James Watt introduced his improved design for a steam engine—a technology whose impact was initially insignificant but which would become a major source of carbon dioxide emissions over the following century by making coal into an economically and technologically viable source of motive power.[61] Since then, they argued, humanity's global impact had accelerated to the point that ensuring the "sustainability of ecosystems against human induced stresses will be one of the great future tasks of mankind."[62]

The Anthropocene was not an entirely novel concept, as Crutzen and Stoermer were quick to admit. Its antecedents included some of the arguments in US diplomat George Perkins Marsh's 1864 book *Man and Nature*, the Italian geologist Antonio Stoppani's 1873 proposal for an "anthropozoic era," and the concepts of biosphere and noösphere developed by Vladimir Vernadskii in the 1920s.[63] Nor had these earlier insights been forgotten in the intervening years. The theme of a human-dominated planet had been repeatedly sounded in more recent discussions of global environmental problems, including those surrounding the 1972 UN Conference on the Human Environment in Stockholm. Maurice Strong, for example, the Canadian oil executive and environmentalist who served as the Stockholm conference's secretary general, argued that the "fundamental question" to be addressed was "how man is to manage the first planet-wide technological civilization in history."[64] Similarly, Ward and

Dubos's report on the conference, *Only One Earth*, focused on "the continued ability of the earth to support future human civilizations," while the 1987 report *Our Common Future* assumed that "the relationship between the human world and the planet that sustains it" had been profoundly changed in the twentieth century.[65] The idea that climate change in particular had transformed the world in ways that threatened the survival of human civilization had also been discussed in a number of works, including one of the first popular accounts of global warming, Bill McKibben's 1989 *The End of Nature*.[66]

Some of Stoermer and Crutzen's key insights and concerns were therefore already both longstanding and widely discussed by the time they introduced the term *Anthropocene*. Nonetheless, the discussion that followed had several distinctive features. One was that rather than being limited to environmentalist circles, the concept received the endorsement of a number of prominent scientists and scientific organizations. It was no coincidence that the newsletter in which Stoermer and Crutzen first proposed the Anthropocene focused on the IGBP's recent decision to shift from the basic science of the Earth system to an integrated understanding of human-driven global change—a shift for which the concept of the Anthropocene was tailor-made.[67] The concept also benefited from having influential advocates. Crutzen, for example, was among the world's leading atmospheric scientists, having been awarded the Nobel Prize in Chemistry in 1995 for his work on the depletion of the ozone layer in the 1970s and '80s.[68] When he restated his and Stoermer's arguments for the Anthropocene in the prestigious journal *Nature* in 2002, he lent the concept instant scientific credibility, even though the geologists responsible for developing a standardized nomenclature of Earth's geological periods had yet to begin discussing its formal adoption.[69] In 2004, the concept's scientific credibility was further enhanced when it was used in *Global Change and the Earth System: A Planet under Pressure*, an IGBP report that synthesized the program's findings and charted its path forward.[70] Like Crutzen and Stoermer, the report's authors, led by IGBP executive director Will Steffen, argued that human activity had profoundly transformed "Earth's environment" since the late eighteenth century, when new technologies and modes of production had inaugurated a period of exponentially increasing consumption of fossil fuels.[71]

Beyond its endorsement by influential scientists such as Crutzen and Steffen and international research initiatives such as the IGBP, there were other critical differences between the concept of the Anthropocene and earlier ways of indicating humanity's power to reshape the world. In particular, the concept was closely connected to Earth system science and the IGBP, which

offered a way of understanding Earth as a dynamic system that was capable
of shifting rapidly from one state of temporary equilibrium to another in response to seemingly small changes in the organization of the system or the
inputs it received from its environment. As climate scientist Wallace Broecker
famously put it, studies of the global climate over the course of the planet's
history had revealed that "far from being self-stabilizing, the Earth's climate
system is an ornery beast which overreacts even to small nudges."[72] The
burning of fossil fuels, advocates of the Anthropocene idea argued, was just
such a "nudge." By altering the carbon cycle that had kept the global climate
system in some semblance of stability over the course of the Holocene, fossil fuel–based industrialization had catapulted the entire Earth system into a
new state. Unlike earlier discussions of humanity's impact on the planet, from
Marsh's *Man and Nature* to Ward and Dubos's *Only One Earth*, the concept
of the Anthropocene implied that such changes could be sudden, total, and
irreversible.[73]

At the same time that advocates of the concept of the Anthropocene embraced a new understanding of the environment, they also proposed a new
kind of entity for whom this environment mattered—namely, "human civilization" or the "human enterprise," which they argued had flourished under
the conditions of the Holocene but would probably struggle to survive in the
Anthropocene. This idea was central to the work of Steffen, for example, a
biogeochemist who became one of the most prominent advocates of the concept following the publication of the 2004 IGBP report. Like Crutzen, Steffen
saw the advent of the Anthropocene as a dangerous rupture in Earth's history
that threatened to lead to the collapse of "planetary life-support systems" if
not managed properly.[74] While that collapse would inevitably affect many
nonhuman forms of life, its significance lay primarily in its effects on "the
natural envelope of environmental variability that provides the conditions
for human life on the planet."[75] The idea of a set of Holocene conditions that
needed to be maintained for the sake of the continuation of the human enterprise was also central to the idea of "planetary boundaries" introduced in
2009 by the Swedish sustainability scientist Johan Rockström and a group of
prominent global change scientists that included Crutzen and Steffen.[76] Only
if humanity could manage to keep greenhouse gas emissions, biodiversity
loss, and other impacts within certain limits, they argued, would the Earth
system continue to provide a supportive "'planetary playing field' for the
human enterprise."[77]

The concept of the Anthropocene also suggested the need for a dramatic change in the nature of environmental expertise.[78] If the impact of human civilization on the Earth system rivaled the forces of nature, as advocates of the concept argued, then any attempt to keep that system within the "planetary boundaries" would demand expertise in human civilization as much as expertise in geophysics, climate science, and other areas of the physical earth sciences. Since human culture was now a key factor in the development of the Earth system, the work of artists, writers, and filmmakers would have to be considered alongside ocean currents and weathering processes. At the same time, scholars of literature, art, history, sociology, anthropology, economics, and other disciplines in the humanities and social sciences would need to understand not just how human societies functioned but also how the Earth system was changing. Climate change, biodiversity loss, and global biogeochemical cycles would have to be taken into account in their studies of nations, markets, cultures, and novels. As the historian Dipesh Chakrabarty has argued, the Anthropocene threatened to render obsolete "the age-old humanist distinction between natural history and human history."[79] With the survival of human civilization dependent on an Earth system that had changed radically in the past and was in the process of changing radically once again, old disciplinary walls could no longer stand. Henceforth human civilization and the Earth system would have to be studied together.

By investigating the debates around the Anthropocene that have unfolded over the past two decades, we can see how the introduction of a compelling and seemingly all-encompassing new concept did not render older concepts obsolete, even if it did require them to be articulated in new ways. Indeed, rather than abandoning the concept of environment, advocates of the Anthropocene idea continued to give it a prominent place, even as they redefined it as the set of planetary conditions that were necessary to the continued expansion of the human enterprise. The 2004 IGBP report, for example, explained the Anthropocene in terms of pervasive changes to what it called "Earth's environment"—that is, the conditions provided by the Earth system for the survival of life in general and of human civilization in particular. This was a variation of the concept of environment that would likely have been alien to some of the earlier users of the term, including the universalist environmentalists of the 1960s and '70s, for whom the fate of human civilization per se was of less concern than the health and well-being of individuals and communities. Nonetheless, the theorists of the Anthropocene were building on and responding to such previous articulations of the concept of environment in order to fashion a distinctive environmentalism of their own.

FROM ONE GLOBAL ENVIRONMENT TO MANY

Going into the 2009 Copenhagen conference, activists had high hopes that it would result in global limits on greenhouse gas emissions that would significantly slow or even reverse the progress of climate change. They were encouraged in those hopes by several auspicious developments, including the awarding of the Nobel Peace Prize to the IPCC and to Al Gore for his film *The Inconvenient Truth* in 2007 and the success of ambitious new nongovernmental organizations such as 350.org in raising global concern about climate change. In the event, however, the Copenhagen conference resulted only in a last-minute, nonbinding accord between the United States, China, India, Brazil, and South Africa that was "taken note of" but not officially approved by the conference as a whole and that was widely denounced by activists as inadequate.[80] As the conference came to an end, observers almost universally declared it a failure.[81] In its wake, climate activists retrenched and, in some cases, gave in to despair. In books with titles such as *Requiem for a Species* and *Reason in a Dark Time*, they argued that the battle to prevent catastrophic climate change had been lost and that the most humanity could now hope for was a gentle descent.[82] Even McKibben, whose 350.org continued to push for drastic cuts to carbon dioxide emissions, contributed to this pessimistic turn with his 2010 book *Eaarth: Making a Life on a Tough New Planet*. "Insofar as our goal was to preserve the world we were born into," he wrote, 350.org and its allies had clearly failed.[83] Climate activists should continue to seek reductions in greenhouse gas emissions since even minor improvements would reduce future suffering, he argued, but no matter how successful they were, it was too late to prevent the familiar planet Earth from being transformed into a strange and inhospitable "Eaarth."

The failure of the Copenhagen conference amplified shifts that had already begun in previous years as the ratification process for Kyoto dragged on and its provisions came to seem more and more inadequate. Among these shifts was an growing willingness among activists to discuss climate "adaptation" as well as climate "mitigation"—that is, preparing to cope with the inevitable effects of climate change even while continuing to seek to prevent the worst outcomes by reducing emissions. Before Copenhagen, many activists had regarded talk of adaptation as a dangerous distraction from efforts to reduce emissions, and moreover a distraction that was often promoted in bad faith by opponents of any type of climate regulation whatsoever. After Copenhagen, even the most resolute advocates of reducing emissions began to accept the necessity of preparing for some amount of global warming. While some argued that the negative consequences of rising temperatures and carbon dioxide levels could be avoided through such simple technological fixes such as building sea walls or adopting drought- and heat-tolerant varieties of staple crops, others argued that more fundamental social and economic adjustments would be necessary. McKibben, for instance, argued that the people most likely to thrive on "Eaarth" were those who strengthened the environments and communities closest to them by getting to know their neighbors, becoming active in government and civil society, consuming food and other essential goods that had been produced locally, and minimizing waste and pollution. Doing so would not prevent climate change–related hardships entirely, he argued, but it would allow life to continue and perhaps even in some ways to improve, as people built more "resilient" communities and ecosystems.[84]

In addition to raising the profile of climate adaptation, the failure of negotiations in Copenhagen also added momentum to a growing interest in "geoengineering" the global climate system directly by blocking solar radiation, fertilizing the oceans, or extracting and storing atmospheric carbon.[85] Long considered a taboo topic by climate scientists and activists, geoengineering gained a prominent advocate in 2006, when Crutzen published an article dismissing the belief that the UNFCCC process would be able to prevent climate change as a "pious wish" and encouraging research on the injection of reflective sulfate particles into the upper atmosphere to reduce the amount of sunlight reaching Earth's surface.[86] Even if carbon dioxide emissions continued to rise, he argued, the resulting reduction in solar energy would keep the planet from warming. Highly controversial when Crutzen proposed it, the idea became increasingly normalized in subsequent years. In the United Kingdom, for example, a 2009 report by the Royal Society made a similar argument about the failure of the UNFCCC process, noting that "global efforts to reduce emissions have not yet been sufficiently successful to provide confidence that the reductions needed to avoid dangerous climate change will be achieved."[87] It was time, the report argued, to begin studies of geoengineering's feasibility even if the ultimate wisdom or necessity of deploying it on a scale large enough to change the climate remained in dispute. In 2010, a team of researchers in the United Kingdom launched the Stratospheric Particle Injection for Climate Engineering (SPICE) project, with the aim of testing precisely the technique for releasing reflective particles into the atmosphere that Crutzen had suggested. Although SPICE was ultimately canceled for reasons that had nothing to do with its scientific value or practical feasibility, the very fact that it was funded and approved reflected a new openness to technological fixes for climate change after the failures of the diplomatic approach taken at Kyoto and Copenhagen.[88]

Even among those who remained focused on reducing greenhouse gas emissions, the failure of the UNFCCC process provoked a change in strategy. For a number of activists, it was clear that the formal structure of UN treaty negotiations and the conceptual underpinnings of global climate activism were no longer sufficient, even if they remained necessary. Both Kyoto and Copenhagen had revealed that negotiations within the framework of the UNFCCC were captive to the interests of powerful individual nations or blocs of nations with aligned interests. Specifically, while the United States remained skeptical of the reality of climate change, let alone the need to take immediate action, the group of developing nations led by China had adamantly resisted constraints on their economic development for the sake of preventing a problem they argued had been caused by the developed nations. Working under these constraints since 1992, the UNFCCC had proven incapable of doing more than convening the world's nations for fruitless discussions. Even more fundamentally, activists grew dubious of the notion that climate change posed a singular challenge to humanity as a whole that could be addressed through a global gathering of nations. The range of variation in responsibility for and vulnerability to climate change was so wide that there were few contexts in which concepts such as "humanity as a whole," "human civilization," or the "human enterprise" had any practical meaning, and the UNFCCC's oft-repeated phrase, "common but differentiated responsibilities," increasingly seemed like an empty formula. All of humanity might be connected to the global climate system, but each part of humanity was clearly connected in its own unique way and at multiple scales of time and space.[89] Climate activism would succeed, many of them concluded, only if it approached these variations as a source of strength rather than as a weakness to be overcome. This same insight was brought to bear on the idea of the Anthropocene, which some critics saw as a concept that ignored important human differences rather than as one that unified humanity to meet a common global threat.[90]

Without abandoning the UNFCCC process or the global scale on which it operated, activists therefore began to concentrate their time and resources on strengthening local responses and building translocal networks. That is, they focused on ways of reducing greenhouse gas emissions and adapting to climate change that were rooted in particular places, such as cities, states, provinces, or tribal territories, but were also bolstered by their connections to similar efforts in other places throughout the world. Protests led by indigenous activists against the construction of oil and gas pipelines in the United States, for

instance, were intensely local in their politics, economics, and cultural commitments, but their success hinged on the fact that they were densely networked
to movements and activists elsewhere, even when the specific nature of the
threats faced by those allies differed.[91] Similarly, residents of Arctic communities threatened by vanishing sea ice and melting permafrost responded not
only by taking action locally but also by establishing alliances with vulnerable
communities elsewhere.[92] Meanwhile, municipalities throughout the developed and developing worlds adopted climate action plans in order to reduce
their contributions to global warming, prepare for the hardships to come, and
learn about strategies and tactics from other municipalities facing similar challenges. In 2015, for example, seventeen large cities in Europe, North America, Japan, and Australia announced their membership in the Carbon Neutral
Cities Alliance, which aimed to reduce their cities' emissions by at least 80 percent by 2050 regardless of what was decided by their national governments or
negotiated at the next Conference of the Parties to the UNFCCC.[93]

Even UNFCCC negotiations focused less and less on what had long been
their primary aim: creating a governance framework capable of regulating the
relationship between the climate system and the human enterprise, both understood as singular and global. This reorientation was evident at COP21, the
twenty-first Conference of the Parties to the UNFCCC, which was held in
Paris in December 2015. By the end of the conference, negotiators had agreed
to aim for an increase in global temperature of no more than 2 degrees Celsius,
and ideally 1.5 degrees if possible, with each nation committing to pursue an
emissions reduction plan suitable to its own needs and capacities. When the
Paris Agreement was announced, it was hailed as the first real progress on
climate change since Kyoto. In the words of one *New York Times* reporter, it
represented a "historic breakthrough on an issue that has foiled decades of
international efforts to address climate change."[94] The apparent success of
COP21, however, masked a significant change in aims and methods, and in
some ways a real retreat. Rather than establishing common standards for the
world's nations to follow, the Paris Agreement allowed each nation to set its
own Nationally Determined Contributions (NDCs).[95] Moreover, despite the
creation of a new system for monitoring progress toward those NDCs, the
agreement lacked any significant enforcement mechanisms. In this sense, while
it did establish a framework for facilitating and verifying climate action on multiple scales, it abandoned the kind of binding global commitments that many
climate activists had long demanded. Whether the promises of the Paris Agreement would be fulfilled depended on the voluntary decisions of individual nations in the years to come, not on the collective action of humanity as a whole.

Climate scientists also reconsidered the strategy that had led them to found the IPCC in 1988—namely, the strategy of building scientific consensus in the hope that it would drive rational policy decisions. To make this strategy effective, they had believed, it was necessary to maintain a certain distance from the political process that would ultimate determine how their findings were used. Bolin, for example, the IPCC's first chair, argued that it was "essential that the different roles of the scientific community and the political institutions were kept apart."[96] That separation became difficult to maintain as the urgency of the climate threat grew and as the UNFCCC process failed to produce significant results. In response, some scientists and scientific organizations abandoned the pretense that they were passionate about scientific discovery alone, instead admitting their research was motivated by a concern with the fate of humanity on a changing planet. As the IGBP prepared for its target end date of 2015, for example, it joined three other major Earth system science research initiatives to lay the groundwork for a new initiative called Future Earth, which sought to place sustainability and equity at the foundation of scientific inquiry rather than seeing them as issues to be addressed only after scientific consensus had been firmly established.[97] In other words, the plans for Future Earth intentionally blurred the distinction between science and policy that had been critical to the initial framing of the IGBP. Even the IPCC, which remained focused on expert analyses of the present and future condition of the climate system, increasingly sought to account for both the politics of knowledge production and the social implications of its findings. In 2018, for example, its special report on the possibility of keeping global warming below 1.5 degrees was prominently framed in terms of sustainable development and the eradication of poverty.[98]

In these developments, we can see how scientists and activists responded to the failure of an approach centered on a single global threat to a singular human enterprise by crafting an alternative variant of the concept of environment that implicitly recognized the inherent multiplicity of both the human enterprise and its global conditions of possibility. Even if every resident of the planet shared the same global climate, they argued, their ways of encountering that climate were often so radically different that there was little of significance that one could say that would apply all of them. For some agricultural communities, for example, the acceleration of plant growth due to higher levels of carbon dioxide would be the most significant consequence of humanity's use of fossil fuels, possibly even a beneficial one; meanwhile, rising temperatures, rising waters, or the increasing variability and unpredictability of weather patterns would have transformative impacts on the lives and livelihoods of other communities. Similarly, while reducing greenhouse gas emissions might be the most effective response for some and geoengineering might be more appealing to others, certain communities would undoubtedly decide that their needs were best served by "business as usual." In short, every community would be affected by changes to the Earth system, but there was no single impact that was common to them all, nor any single policy that would address all their needs. In this sense, one might even say that there were multiple global environments, each defined in relation to a particular "human enterprise."[99]

*

At several points in the First Assessment Report published by the IPCC in 1990, the scientists who had authored it warned that climate change was "potentially the greatest global environmental challenge facing humankind," echoing language that dated back to the 1972 UN Conference on the Human Environment, if not earlier.[100] They might also have said that it was greatest challenge ever faced by climate scientists, who found themselves in need of methods for predicting changes in the enormously complex Earth system for decades or centuries to come, even as human activities were changing the system in front of their eyes. Only by developing such methods would they be able, they believed, to diagnose the nature and severity of the threat and to present a convincing plan of action to policymakers and to the public. From a certain perspective, then, the history of climate change as a public issue tracks closely with the development of a global community of scientists and the methods they used to make the "global climate system" into an object of knowledge and concern.

Although this community and set of methods did not emerge overnight, they were also not merely continuations of older efforts. Over the course of the Cold War period, as technological, geopolitical, and economic processes tied the world's peoples closer together and accelerated the human impact on the planet, climate scientists painstakingly built the elements of the system that would eventually make possible something like the IPCC's First Assessment Report—that is, a set of authoritative claims about the state of the global climate that was endorsed by a global community of scientists. This system consisted of various components, including the satellites, weather stations, and human observers that gathered standardized data, the computer models that transformed the data into simulations of the past, present, and future of the climate system, and institutions like the IPCC, where scientists collectively identified a set of claims they believed could credibly be supported by those data and models. It was in these diverse ways that climate scientists materialized, for themselves if not for humanity as a whole, the kind of planetary community that Ward and Dubos had called for in 1972.

If creating a planetary community of climate scientists was difficult, extending that community to humanity as a whole proved to be virtually impossible. As we can see in the history of climate change activism, it was much easier to assert the existence or necessity of a planetary community than it was to bring such a community into being. In the case of the UNFCCC, the assumption that the negotiators at Kyoto, Copenhagen, Paris, or the many meetings in between could be treated as unproblematic representatives of their nations was quickly shown to be unfounded. The United States, for example, repeatedly negotiated agreements that subsequently failed to be ratified by the US Congress or to be supported by the majority of the American people. Moreover, even when negotiators could be treated as legitimate representatives of their respective nations, it was clear that the United Nations lacked the moral or legal authority to create a global community out of a fractious collection of independent nation-states and blocs of nation-states with shared interests. Even the transnational activism of a group like 350.org could go only so far to create a genuine sense that all the world's peoples were or should be engaged in a common project.

By examining climate scientists' and activists' adoption and adaptation of the concept of environment during this period, we can see how its meaning shifted in relation to a new master term, *Anthropocene*. In some respects, the concept of the Anthropocene threatened to displace the concept of the environment from its reigning position. By invoking it, scientists, activists, artists, and scholars were able to point to many of the concerns that had animated environmental activism without ever uttering the word *environment*—at least, in principle. In practice, however, both the concept of environment and the word itself continued to play a critical role in discussions of the Anthropocene, as Crutzen and Stoermer's use of "environment" and "environmental" in their original 2000 article on the Anthropocene and the 2004 IGBP report's discussion of the "Earth's environment" both suggest. What changed was not the utility of the concept of environment, but how it was measured and materialized and what kind of entity it was defined in relation to—in this case, an abstraction called the "human enterprise" or "human civilization" that proved easier to imagine than to mobilize against the threat of climate change.

When we situate climate change activism in relation to the long history of environmental thought, it appears as one of the most ambitious and perhaps most tragic attempts to forge a sense of solidarity among diverse people by showing them how their lives and livelihoods depend on a shared set of external conditions that they themselves are responsible for maintaining in good condition. As previous chapters have shown, such efforts have often succeeded, and those successes are one of the reasons that the concept of environment has continued to be adopted and adapted to new circumstances and aims. In the case of the global environment, however, the social and material obstacles to establishing such a sense of solidarity have been enormous—even greater, in fact, than those faced by the universalist environmentalism of the environmental movement of the 1960s and '70s, which ultimately did not depend for its success on unifying humanity on a global scale. The failure of the concept of environment to function well when expanded to the scale of planet Earth is one of the reasons why a small but growing number of people have begun to question whether the concept is still useful in our changing times. This is the question to which we turn in the conclusion.

What Might the
Environment Become?

Over the past two centuries or so, from naturalists' first inklings of the idea that they could productively study life in terms of "organisms" and their "environments" to the latest campaigns by climate activists to protect the "global environment" of the "human enterprise," the concept of the environment has been enormously generative. Among scientists, it has inspired the development of a wide range of instruments and research practices that have produced groundbreaking discoveries into how living beings are organized in relation to their surroundings. More broadly, it has been used to call attention to the material conditions of existence and has served as a reminder that no organism or community can survive in isolation. It has suggested new ways of improving the health of human individuals and populations, as well as new techniques for managing resources in ways that lead to prosperity and security. It has helped reveal the toxic risks hidden among the consumer abundance produced by industrial economies. It has been used to establish political solidarity among diverse groups of people, as well as to identify the distinctive inequities and injustices faced by each particular group. Among historians and other scholars in the humanities and social sciences, it has provided a reminder that human meaning and experience emerge in a material world. Perhaps most importantly, it has provided a framework for people to recognize their common dependencies on one another and on their shared surroundings and to join together to sustain and improve them.

There is therefore ample reason to celebrate the history of environmental thought. Nonetheless, as one can plainly see in each of the episodes described in this book, the concept of environment has been deployed in ways that were already being criticized during their own times and that with hindsight seem unquestionably counterproductive, or even unjust. On the most general level, by drawing a bright line between living beings and their surroundings while emphasizing the necessary connections between them, the concept paradoxically made it easier for some people to argue that those surroundings could be exploited without restraint, so long as it was in the service of human needs. In the context of nineteenth-century European imperialism, moreover, the concept of environment was used to support claims about deep-seated racial differences and to justify public health and sanitation measures that reinforced the hierarchical structures of empire. In industrializing cities, it was used to frame social reform programs that, despite the democratic impulses of leaders such as Jane Addams, often veered into condescension and exacerbated social inequalities. Coupled to skepticism regarding the ability of ordinary citizens to understand the "real environment" around them, it could even be turned against democracy itself. Between the two world wars, technocrats from the left and the right embraced the concept of environment to justify centralized, top-down, expert control of the resources of their nations and of the globe. As it became the conceptual core of a new social movement in the second half of the twentieth century, it paradoxically tempted many to a kind of environmental universalism—that is, toward assuming that all people have or desire the same kinds of relations to their surroundings. Finally, adapted to the global challenge of climate change at the end of the twentieth century, the concept of environment helped experts give the impression that they were speaking on behalf of all humanity even as they ignored the specific needs and vulnerabilities of particular human groups.

If many of these old modes of thinking and acting environmentally were already problematic in their own times, they have become only more so today. One reason is that the world itself has materially changed in ways that trouble some of the assumptions on which environmental thought has been based across its many variations. The rapidly advancing science and technology of gene modification, for example, has made it easier to produce organisms whose relationship to their respective environments is not the result of adaptation but rather one of intentional human design. At the same time, as the concept of the Anthropocene seeks to capture, the environments in which those organisms live are increasingly being shaped by human action on scales ranging from the molecular to the planetary. When both organisms and their surroundings are at least partially products of human design (or even human whim), it becomes harder for scientists to identify the kinds of natural regularities in the relationships between them that have been the focus of environmental thought over the past two centuries. Meanwhile, the intellectual, institutional, and economic frameworks that long made it seem both reasonable and productive to divide the world into individuals and their environments are shifting unpredictably, blurring old boundaries and destabilizing old certainties. As it becomes less and less clear who "we" are, it becomes harder and harder to say anything meaningful about "our" environment. It is not unreasonable to conclude that these growing complexities and uncertainties make the concept of environment less compelling than it once was, not merely because we better understand its flaws but also because those flaws are more troubling in today's changed world than they once were.

The troubled history and uncertain future of the concept of environment help explain the appeal of the critiques of "environment" that have been offered in recent years, including those by Bruno Latour and Donna Haraway mentioned at the beginning of this book. Both of these scholars have suggested—albeit in the form of questions rather than declarations—that it might be time to retire the concept of environment in favor of alternatives that are better suited to the contemporary moment. Although their critiques of environment are new, the foundations of those critiques are longstanding. Since the 1980s, for example, Latour and other contributors to the research program known as Actor-Network-Theory have criticized models of society and nature premised on the idea that certain entities are capable of action while others are only capable of being acted upon—a kind of division of the world that is common to many, though not all, forms of environmental thought.[1] When Latour writes in his recent book *Down to Earth* that Earth should be seen as "an agent that participates fully in public life" rather than as "the milieu or the background of human action," we can hear echoes of his more general critique of any approach that designates large swaths of the world as conditions for action rather than potential sources of action.[2] For her part, Haraway has long critiqued the ideal of the autonomous individual subject that is central to Enlightenment liberalism and much of contemporary politics by showing how it obscures various forms of embodied labor and interdependence in ways that bolster patriarchy, colonialism, racism, and capitalism.[3] In her recent *Staying with the Trouble*, she suggests that we ought to think of life not in terms of relationships between organisms and their environments but rather in terms of what she calls "sympoiesis"—that is, a process by which entities are continuously being reproduced through their entanglements with each other rather than through their efforts to maintain their autonomy.[4]

Latour and Haraway are just two of a number of voices that have emerged in recent years to critique the concept of environment and to suggest alternatives that do not involve dividing the world into "organisms" and "environments," however those may be conceived. Emphasizing the porosity and permeability of living bodies and the often surprisingly lively properties of nonliving entities, scholars such as Tim Ingold, Stacy Alaimo, Timothy Morton, and Jane Bennett have proposed concepts such as "trans-corporeality," "the meshwork" (or simply "the mesh"), and "vital materialism" to characterize a world that cannot easily be divided into autonomous beings and their conditions of existence.[5] While each of these scholars offers an analysis with its own distinctive nuances, they all share the conviction that the world consists of complex entanglements among living and nonliving entities and that it is therefore a mistake to designate one entity as the center of action and interest while reducing the rest to its "environment." By focusing on these complex entanglements, they hope to provide the foundation for more just and effective ways of encountering our surroundings. As of yet, none of these alternatives has been widely adopted, and considering the scholarly language in which they have been expressed, it seems unlikely that they will be anytime soon (although given that *environment* also began its modern career as an obscure technical term, it is not unthinkable that some of these terms may eventually come to seem just as transparent and indispensable). Nonetheless, they show that it is possible to speak compellingly about many of the things that concern environmentalists without referencing "environments" at all, let alone "the environment" of contemporary environmentalism.

Is it time, then, to abandon the concept of environment in favor of one of these alternatives, or perhaps another of our own devising? Given that scholars concerned with transcorporeality, the meshwork, and so forth seem to be just as concerned about toxic pollutants, biodiversity loss, and climate change as the most conventional environmentalists are—and in many cases offer much the same set of solutions—it might be tempting to dismiss such discussions as merely "academic" in the worst sense of the term and to continue using "environment" as a handy if imprecise label for a common set of interests. Indeed, some scholars in the environmental humanities have argued that we should do just that. If the choice of concepts is, however, as this book has argued, materially consequential—that is, if it influences not only how we think, talk, and write but also who we ally ourselves with and against and what we do and make together—then a critical examination of the concepts we choose to use remains essential. One potential consequence of such an examination is that we may discover that a once-essential concept has become untenable as our circumstances and aims have changed. For this reason, it seems worth taking seriously the critiques of the concept of environment and the alternatives that have been offered. The kind of "environmentalism" we seek to build, or not to build, will depend on what we find.

As the previous chapters have shown, the question of whether a concept is useful and appropriate to a given set of circumstances and aims cannot be answered in the abstract, whether we are considering the past, present, or future. For both principled and practical reasons, the concluding pages of this book therefore offer neither a general prescription for how our concept of environment should change nor even a definitive conclusion about whether it should be retained or abandoned. There are too many variations on the concept and too many circumstances and aims in relation to which they would have to be evaluated for such a universal answer to be possible or desirable. Instead, the following vignettes suggest some of the ways in which "environment" is actually being reworked under specific circumstances, for specific purposes, and by specific groups of people today. Each of these ways is distinctive, but all of them have in common the explicit use of the term *environment* as well as some kind of relationship, however conflicted, with the histories traced out in this book. None of them depict the borders between bodies and their surroundings as rigid or impermeable, hitch their fates to those of expanding empires, reinforce racial distinctions or other social hierarchies, divide the physical environment from the social or cultural environments, or attempt to impose a single temporal or spatial scale of analysis, global or otherwise. Beyond these similarities, they are wide-ranging, diverse, incomplete, and sometimes contradictory—characteristics that suggest not only that the history of the concept of environment is still being written, but that we may be living through one of its most complex and generative moments yet.

ENVIRONMENTAL SCIENCE IN THE CRITICAL ZONE

In 2007, a multidisciplinary group of scientists at Penn State University joined with researchers from ten other institutions to launch a new research initiative focused on a small water catchment not far from the university's campus in central Pennsylvania. Since then, the Shale Hills Critical Zone Observatory, as the initiative is called, has sought to bring the expertise of geologists, hydrologists, meteorologists, soil scientists, and plant physiologists to bear on a question with broad-ranging implications—namely, how the site's soils have been produced and transformed over time by geochemical, hydrological, biological, and geomorphological processes, including climate change since the beginning of the Holocene and forest clearing associated with European colonization.[6] To answer this question, they have developed a system of sensors and sampling protocols capable of generating a detailed record of the site's processes, from precipitation levels and wind speeds to the rates at which sap flows through different species of trees to changes in the dissolved solids in groundwater.[7] Back in the laboratory, they have combined these data with high-resolution three-dimensional maps of the land surface to develop computational models of the flow of water through the system and its effects on soil erosion. Over the years since its founding, as the initiative has grown both in the number of researchers involved and the extent of the field site, the initiative's ambitions have expanded even further. Beyond simply characterizing the current state of the catchment, it has also sought to use these models to "earthcast" future changes likely to result from human activities.[8]

The research at Shale Hills is part of a broader research network organized around the concept of the "critical zone," first elaborated in a 2001 report on research opportunities in the Earth sciences issued by the US National Research Council.[9] The report identified a need for interdisciplinary research on "the heterogeneous, near-surface environment in which complex interactions involving rock, soil, water, air, and living organisms regulate the natural habitat and determine the availability of life-sustaining resources."[10] Following the NRC report, the National Science Foundation (NSF) supported a series of workshops that led to the foundation of the Critical Zone Exploration Network in 2006, a website and online community that helped an international community of interdisciplinary Earth scientists coalesce around the concept.[11] The first major funding for critical zone observatories (CZOs) came in 2007, when NSF awarded multiyear, multimillion-dollar grants to research collaborations—one to the Shale Hills CZO, another to a site on the eastern edge of the Rocky Mountains in Colorado, and the third to a site on the southern edge of California's Sierra Nevada mountain range.[12] In subsequent years, as the US network of CZOs continued to expand, the concept gained traction internationally, with new CZOs planned or established in Germany, Australia, China, and elsewhere.[13] In many cases, these CZOs were based on already existing research sites, but they sought to study those sites in new ways. In 2015, for example, the French government launched OZCAR, a critical zone network incorporating twenty-one existing sites into a new research infrastructure.[14] The new scientific subfield has even come to the attention of social scientists; Bruno Latour, for example, has identified it as one of the most promising ways of studying Earth in the Anthropocene.[15]

The growth of the CZO network reflects the appeal of a new way of organizing research across scientific disciplines and scales of time and space. On one hand, CZOs resemble other kinds of well-established field sites for

FIGURE 14. An illustration of a heavily instrumented critical zone observatory site. (Reprinted from "NSF Awards Grants for Four New Critical Zone Observatories to Study Earth Surface Processes," National Science Foundation, News Release 14-008, n.d., accessed May 3, 2019, https://www.nsf.gov/news/news_images.jsp?cntn_id=130115&org=NSF.)

research in ecology, geomorphology, agronomy, forestry, and soil science—and indeed, they are often developed at or near research sites of these kinds, where they can take advantage of existing infrastructure and data sets. In the United States, for example, many CZOs are colocated with or adjacent to sites within the Long-Term Ecological Research (LTER) network that was established in the early 1980s, in part in response to the failures and lessons of the US participation in the International Biological Programme.[16] CZOs also share with climate science and Earth system science an interest in global processes, and specifically with anthropogenic impacts on biogeochemical cycles. On the other hand, critical zone science is distinctive in its attempt to produce multidisciplinary, multitemporal understandings of single, well-defined sites, which are often defined in terms of watersheds. Unlike most LTER sites, for example, CZOs focus as much on physical processes such as weathering, erosion, and geological and climatological changes spanning thousands or even millions of years as they do on living organisms and their relationships over minutes, days, or years. The result is a form of scientific research that is inherently multidisciplinary, expansive in its temporal scope, and conscious of global processes, but also intentionally limited to the distinctive dynamics of a particular site.

This new way of organizing scientific research around the multidisciplinary study of physical processes at a single site has implications for the way critical zone scientists define and deploy the concept of environment. To some extent, the concept of the critical zone has displaced the concept of environment, such that when critical zone scientists describe their work, they sometimes use the term *critical zone* in places where one might expect to find *environment* instead. Rather than abandoning the concept of environment entirely, however, critical zone scientists have explicitly claimed to be contributing to our collective store of environmental knowledge, albeit in new and distinctive ways. For example, critical zone scientists tend to speak and write about "environments" in the plural, as befits their focus on in-depth studies of physical and biological processes at particular sites; according to a 2014 NSF pamphlet, the aim of the CZO program is to establish "observatories in many environments to study Earth's outer skin."[17] The site-based character of critical zone science also changes the way it represents the relationship between environments and the things they surround. Rather than focusing on the environment of a particular organism or community, as ecologists have traditionally done, it considers the conditions for survival and flourishing of all the forms of life, including humans, that reside within a particular watershed or other clearly defined site. In this way, critical zone scientists have fashioned a variation on the concept of environment that is suited to a research program as vast in its temporal scope as Earth system science and as rooted in place as soil science.

By situating critical zone science in relation to the history of environmental thought that has been traced in this book, we can see how it continues many of the same questions and approaches of earlier scientific "environmentalists"—that is, people who organized their research around the concept of environment—even as it challenges some of their key premises. Indeed, critical zone scientists themselves argue that they are recovering and revitalizing older visions of environmental science, particularly Vladimir Vernadskii's biogeochemical vision of the biosphere and the ecosystem ecology of Raymond Lindeman, G. Evelyn Hutchinson, and the Odum brothers that it helped inspire.[18] At the same time, it differs significantly from those earlier approaches. In comparison to ecosystem ecology, for example, critical zone science is more attentive to physical processes, long-term geological and climatological changes, and human activities as integral parts of the systems under study rather than as external "inputs." By placing it alongside other environments and environmentalisms of the past, we can also see more clearly some of its limits and constraints. Because it focuses on the interface between land, air, and water, for example, it has thus far had little to say about oceanic environments. In this and other ways, it is clear that critical zone science is unlikely to provide a model applicable to life in all forms and under all conditions, even if it does transform our understanding of terrestrial life. Finally, situating it in relation to the history of environmental thought invites us to consider the material relations and broader environments in which it operates and to ask what changes would be necessary—both to those conditions and to critical zone science itself—to make its versions of environment and of environmentalism thrive more broadly.

THE INDIGENOUS ENVIRONMENTALISM OF THE WATER PROTECTORS

In April 2016, a group of indigenous activists led by LaDonna Brave Bull Allard set up a camp near the confluence of the Missouri River and its tributary Cannonball River, just outside the northern border of the Standing Rock Reservation. The aim of the camp was to stop the construction of the Dakota Access Pipeline (DAPL), the project of a company called Energy Transfer Partners. Since 2014, the company had been planning to construct a 1,172-mile-long, 30-inch-diameter pipeline from the Bakken oil fields in North Dakota to a tank farm in Illinois, where the crude oil would be stored before being shipped onward to refineries to the south and east.[19] By early 2016, the company had obtained all the necessary permits from the various state governments involved; the only regulatory hurdle it still faced was a set of easements from the Army Corps of Engineers that were necessary for the pipeline to cross under Lake Oahe, a reservoir on the Missouri River that stretches more than 230 miles from central South Dakota into North Dakota—much of it along the border of the Standing Rock Reservation. The protest at the Sacred Stone Camp was the culmination of several years of growing frustration among the Standing Rock Sioux over the failure of the company and state and government regulators to take their concerns seriously. Over the course of the spring and summer of 2016, thousands of people joined them at several nearby camps, eventually constituting the largest single indigenous-led protest movement in recent US history.

The protests at Sacred Stone Camp, at the much larger Oceti Sakowin Camp, and at various points along the pipeline sought to stop construction through direct action and to bring pressure to bear on the US government to withhold the necessary easements. With their slogan "Water is life" (or *mni wiconi* in the Lakota language), the self-described "water protectors" asserted the importance of water over the interests of private corporations and the regulators and law enforcement agencies that supported them. For many of the Standing Rock Sioux and other tribes located along the Missouri River, the "black snake" of DAPL represented an existential threat to their water, their sacred sites, and their treaty-guaranteed land rights. Their fight was eminently local, in that it was entangled both with the practical importance of the Missouri River (Mni Sose) to the Standing Rock Reservation and with its experience of repeated betrayals by the US government, beginning with the unilateral abrogation of mid-nineteenth-century treaties signed with the Oceti Sakowin, the Seven Council Fires of the Great Sioux Nation.[20] At the same time, the #NoDAPL movement quickly gained a number of other allies representing a wide range of concerns, from violations of indigenous rights in other parts of the world to global climate change to racial and economic injustices. These distant allies were mobilized in part through indigenous activists' use of social media to circulate images and ideas and to build networks that extended globally even as they remained focused on their own concerns and histories. Early in December 2016, the effectiveness of this translocal strategy seemed to be confirmed when President Barack Obama ordered the Army Corps of Engineers to prepare a full environmental impact statement before issuing a permit, which effectively postponed the pipeline for years to come. The victory proved short-lived, however, as many of the leaders of #NoDAPL suspected it might be. Within days of being sworn in as president in late January 2017, Donald Trump instructed the corps to expedite its review, which it completed within a few weeks. The remaining sections of the pipeline were finished in April, and as much as half a million barrels of Bakken oil per day soon began to flow under Lake Oahe.[21]

Despite its failure to stop the pipeline, the #NoDAPL campaign was an important milestone in a growing indigenous movement to protect land and water from extractive industries and their allies in government.[22] Concerned with such threats as toxic waste, global warming, and biodiversity loss, the movement shared aims with the environmental justice movement as it had developed since the 1980s (a movement which itself had been powerfully shaped by indigenous activists) but it also differed both in its tactics and in its moral and ontological foundations. The Standing Rock Sioux and other indigenous participants in the #NoDAPL protests staked their moral authority not only on the universal right to a healthy environment but also on their belonging

FIGURE 15. A banner at the Oceti Sakowin Camp on late November 2016 displaying the motto "Mni wiconi" (Water is life). (Photograph by Flickr contributor Becker1999 [Paul and Cathy Becker], accessed May 4, 2019, https://www.flickr.com/photos/becker271/31015368944/, published under a Creative Commons Attribution 2.0 Generic License [CC BY 2.0].)

to a particular place and on the histories of oppression and dispossession that had stripped them of a voice in decisions like the one permitting the pipeline to cross the Missouri. In their slogan "Water is life," the protestors asserted a special relationship between the life of their community and the life of particular lands and waters, which they understood in terms that were spiritual, cultural, and historical as well as biological and physical.[23] Meanwhile, in addition to deploying the forms of mass civil disobedience that had been central to many twentieth-century social movements, including movements for indigenous rights as well as the environmental movement, they also took advantage of social media to construct networks that linked together disparate local concerns across great distances—a tactic that made them more similar to movements like Black Lives Matter than to conventional environmentalism.[24]

Even though many of those who participated in #NoDAPL did not identify themselves as "environmentalists"—and indeed, explicitly sought to distance themselves from the environmental movement—the concept of environment nonetheless continued to play an important role in the protests. To some extent, the framing of the issue as "environmental" was strategic, a response to legal structures that made it one of the most effective ways of challenging projects such as DAPL. Just as environmental justice activists of previous decades had drawn on the expertise and resources of environmental organizations despite their skepticism toward "environmentalism" as a label, the Standing

Rock Sioux turned to the environmental law organization Earthjustice for support in their lawsuit against DAPL and the Army Corps of Engineers, which relied heavily on the National Environmental Policy Act and its requirement for an environmental impact statement. Beyond such legal contexts, however, indigenous activists also adopted and adapted the concept of environment to their own purposes, even as they rejected the stereotype of the "noble savage" or "ecological Indian" that has often been deployed by nonindigenous environmentalists.[25] The Indigenous Environmental Network, for example, which played an important role in organizing the #NoDAPL protests, had been linking environmental concerns to indigenous rights since its founding in 1990.[26] Similarly, speaking with Amy Goodman of *Democracy Now!* soon after Trump's decision to expedite DAPL's approval in early 2017, the chairman of the Standing Rock Sioux Tribe, Dave Archambault II, emphasized the importance of assessing the pipeline in light of its impact on "tribal lands, treaty rights, human rights, the environment for this nation."[27] Reinterpreted in the light of land claims of indigenous nations and of an understanding of land and water as kin rather than as property or resources, the concept of environment helped activists make clear that their struggle to protect water was about ensuring the conditions that would allow them to survive and thrive as a people.

By situating the indigenous environmentalism of recent years in relation to the histories of environments and environmentalisms traced in this book, we can see how it both differs from and builds on what came before. Like the earlier environmental justice movement to which indigenous people had already made significant contributions, it has called attention to the unequal distribution of environmental risks while echoing the claims of universal human vulnerability and human rights advanced by the mainstream environmental movement. Unlike much of the environmental justice movement, however, it has done so not in terms of the classic sociological categories of race, class, and gender but rather in terms of cultural, historical, spiritual, and material connections to particular places and to the histories of settler colonialism that threaten those places and the people who inhabit them.[28] Protecting the environment, from this standpoint, means protecting the relationships of people with deep ties to a particular place, which are distinct from the relationships to the same place that people without those ties might have. As the #NoDAPL campaign showed, this version of environmentalism succeeds best when people who are tied to one place find allies among people whose ties lie elsewhere—and even then it often fails in the face of concerted opposition from powerful corporations and governments. Whether this evolving form of indigenous environmentalism will have more success in the future remains to be seen. Like critical zone science, its future is still in the making, contingent on changes in the world around it and in how it puts its vision into practice.

ENVIRONMENTAL ART FOR THE AEROCENE

In December 2015, as negotiators gathered in Paris for the twenty-first Conference of the Parties to the UNFCCC (COP21), the Argentine artist Tomás Saraceno and his team were installing two giant silvered mylar spheres in the Grand Palais as part of an artistic program that aimed to inspire reflection on global warming.[29] Heated by sunlight shining through the palace's vaulted glass roof, the two spheres rose and fell as the air within them expanded and condensed in response to the changing temperature. Designed as part of a collaboration between Saraceno and scientists at the Massachusetts Institute of Technology, the balloons bridged science, art, and activism by repurposing a technology first used by the French aerospace program in the 1970s to lift atmospheric sensors into the stratosphere. The installation represented a new iteration of Saraceno's longstanding interest in unpowered flight, including work that began in 2007 with Alberto Pesavento and others to create the Museo Aero Solar, a floating "museum" consisting of used plastic bags taped together into a giant solar-powered balloon. As the Museo Aero Solar was exhibited in a series of cities around the world over the succeeding years, it gradually became one component of a more ambitious project, the "Aerocene," which sought to foster a speculative vision of a fossil fuel–free world in which global mobility was powered entirely by wind and sunlight.[30]

Since the installation at the Grand Palais for COP21, the Aerocene project has mobilized the collective expertise and enthusiasm of a global network of artists, scientists, activists, and other participants, now organized under the umbrella of the Aerocene Foundation. Some of these collaborators have worked on improving the design of the balloons themselves, including developing do-it-yourself (or in the project's preferred terminology, "do-it-together") kits that allow anyone with a minimum of skill to assemble and launch a solar balloon or even build one from scratch themselves.[31] Others with expertise in aeronautics and atmospheric science have developed a "float predictor" that models the path that a balloon driven by wind currents at a given altitude would follow—a first step in making it possible for a solar-powered balloon to navigate predictably, if circuitously, from one place on the surface of the globe to another by raising or lowering its altitude to catch the appropriate wind currents. As these speculative projects are developed, artists and activists have taken advantage of existing technologies to launch Aerocene balloons

FIGURE 16. An Aerocene balloon heated by solar reflectors at Tomás Saraceno's *Albedo* installation at Art Basel Miami Beach in 2018. (Courtesy of the Aerocene Foundation, accessed May 4, 2019, http://aerocene.org/albedo, published under a Creative Commons Attribution– ShareAlike 4.0 International License [CC BY-SA 4.0].)

on every continent, including Antarctica. In October 2018, the project set a record for exclusively solar-powered human balloon flight when six volunteers were kept aloft by a tethered balloon for more than an hour at a site southeast of Paris.[32] While the project remains a long way from its utopian vision of a world of fossil-free mobility, it continues, in the words of a manifesto released in 2018, to take steps toward "a new epoch beyond the Anthropocene, towards the decarbonisation of the air, and towards independence from fossil fuels."[33]

While the Aerocene project stands out for its longevity and for the number and diversity of participants it has enrolled, it is just one of a number of recent artistic projects that go beyond merely representing the environmental crisis or reenacting it in miniature—long the most common ways of practicing "environmental art"—to making material contributions to the solution of environmental problems, often in close collaboration with scientists, activists, and laypeople.[34] Such projects respond not only to the widespread perception of an environmental crisis that cannot be solved through consciousness-raising alone but also to the conditions for producing and selling art today, when a global network of museums, galleries, and art fairs has transformed the work of certain celebrity artists into speculative investments for wealthy collectors even while most artists struggle to survive.[35] Facing this dual crisis in the conditions for human well-being and the possibility of socially meaningful art, some artists have embraced a countervailing practice that emphasizes collaborating with experts and affected communities to address issues of pressing concern. Like the Aerocene's solar-powered balloons—which are meant to be sublimely beautiful at the same time that they reduce humanity's greenhouse gas emissions—such art seeks to generate affective and aesthetic effects while also accomplishing various kinds of practical work. Many of these projects link fine art, industrial design, community engagement, and environmental activism in the hope that artistic interventions can have a life beyond the gallery.[36]

Given its embrace of science and technology, its speculative and even playful character, and its vision of an airborne way of life, the Aerocene might be seen as a postenvironmentalist project that addresses issues conventionally understood as "environmental"—including climate change, air pollution, biodiversity loss, and resource scarcity—while rejecting the methods and categories of environmentalism. As Sasha Engelmann, a geographer who has been closely involved with the Aerocene project, has written, it consists of "a series of techniques that occur neither in subjects nor in the environmental milieu, but unfold in the choreographic relations between bodies, materials, devices and sites."[37] In her focus on relations among disparate entities rather than on relations between living beings and their surroundings, Engelmann echoes some of the critiques of the concept of environment that have emerged in recent years, and in fact two of the most prominent authors of those critiques, Tim Ingold and Timothy Morton, have written appreciative essays on the project.[38] At the same time, the concept of the environment, suitably adapted to new circumstances and aims, continues to do useful work for the Aerocene. The home page of the Aerocene website, for example, describes the project as an "interdisciplinary artistic endeavor that seeks to devise new modes of sensitivity, reactivating a common imaginary towards achieving an ethical collaboration with the atmosphere and the environment."[39] Similarly, the *Aerocene Manifesto* calls for building "a less anthropocentric relationship with the environment" and learning to "re/entangle ourselves with the surrounding milieu."[40] The Aerocene project has thus reinvented the concept of environment for its own purposes rather than abandoning it; it is not postenvironmental but instead environmental in a new way.

Situating the Aerocene in the long history of related but distinctive ways of materializing the concept of environment helps us see how certain critiques of the concept of environment that are compelling when applied to its most recent or most dominant variations become misleading when extended to all the diverse ways the concept has been articulated and materialized over its two-century-long history, including the variety of alternative environmentalisms that are emerging today. In fact, some of the commentators on the Aerocene acknowledge as much, even as they continue to advocate for alternative concepts such as the "mesh" or "transcorporeality." Morton's writing on the Aerocene, for example, suggests both that the project points the way to a postenvironmental future and that it shows us that "an environment is a necessarily dynamic, unstable thing that surrounds and penetrates us. We are part of it, and we exceed it. We are it, and we aren't it."[41] In the midst of a challenge to the dominant way of imagining the environment, in other words, is a revitalization of the concept in a novel guise. Of course, whether this vision of a dynamic and unstable environment can become the foundation of a new kind of environmentalism capable of spreading beyond the rarified world of the contemporary art market remains to be seen. Much depends on how the Aerocene project puts its utopian vision into practice and whether it can find allies, human and otherwise, who can help make its commitment to a fossil fuel–free future a reality.

THE ENVIRONMENTAL HUMANITIES IN THE CHANGING UNIVERSITY

Over the past decade or so, something called the "environmental humanities" has rapidly made inroads into teaching and scholarship in the humanities. Even though it is not always clear precisely what that something is, classes have been taught, speaker series have been organized, journals have been founded, essay collections have been published, and major grants have been awarded on the basis of the belief that the environmental humanities represent an important new direction in scholarship—in particular, one that combines traditional modes of research and teaching with efforts to address the environmental crisis that reach beyond the classroom and the library. Building on discussions of the "ecological humanities" already in progress for a number of years, the environmental humanities first took institutional form in the early 2010s.[42] In 2012, for example, a group of scholars in Australia founded a new journal called *Environmental Humanities* in order to "support and further a wide range of conversations on environmental issues in this time of growing awareness of the ecological and social challenges facing all life on earth," while a group of historians of science and technology at Sweden's Royal Institute of Technology (KTH) founded an Environmental Humanities Laboratory with the aim of creating a "post-disciplinary intellectual environment that combines education, research and graduate training in innovative ways and sets knowledge in the humanities into action to favour sustainable development."[43] In the United States, the institutionalization of the environmental humanities accelerated after 2014, when the Andrew W. Mellon Foundation began distributing several million dollars in grants to expand environmental humanities programs at universities.[44]

While many definitions of the environmental humanities have been offered, all of them tend to include several core characteristics.[45] One of those characteristics is multidisciplinarity; the field seeks to build bridges among environmental history, ecocriticism, environmental philosophy, and other subfields and specializations in the humanities that in some cases have existed since the emergence of the modern environmental movement but which have nonetheless rarely been brought into dialogue with one another. Another is an emphasis on bringing experimentation and practical engagement into play alongside the kind of close reading, contextualization, and critical reflection that has traditionally been the focus of scholarship in the humanities. This emphasis on experimentation and practice may be one of the reasons that the image of the humanities "laboratory"—as in the Environmental Humanities Laboratory at KTH or the Laboratory for Environmental Narrative Strategies at the University of California, Los Angeles—is as prominent as it is in the field's rhetoric. The use of "laboratory" language also points to another distinguishing characteristic of the environmental humanities—namely, its focus on fostering collaboration not only among scholars in adjacent disciplines within the humanities, but also between humanities scholars and natural scientists. Finally, most definitions of the environmental humanities include a hefty dose of political engagement, including efforts to ameliorate interlinked social and environmental injustices.

Beyond its strengths as a way of understanding complex environmental issues, the rise of the environmental humanities at this particular historical moment can be attributed to several other factors, including changing ideas of environmental expertise and the shifting place of the humanities within the university system. As the effectiveness of a top-down, science-first approach to climate change and other global environmental issues has come into question, the value of expertise in the humanities has become clearer. The idea of the Anthropocene, for example, with its emphasis on the role of humanity in determining the future of the Earth system, has helped justify bringing the humanities and social sciences into science and policy discussions. As the Swedish historian Sverker Sörlin has argued, "In a world where cultural values, political and religious ideas, and deep-seated human behaviors still rule the way people lead their lives, produce, and consume, the idea of *environmentally relevant knowledge* must change."[46] At the same time, however, scientist-led efforts to take into account the "human dimensions" of global environmental change have often been based on a superficial understanding of scholarship in the humanities (just as scholars in the humanities often have a superficial understanding of the science they draw on).[47] In this context, the institutionalization of the environmental humanities provides an opportunity for humanities scholars themselves to lead the conversation about what they can contribute to addressing the environmental crisis. A second factor in the rise of the environmental humanities is the changing place of the humanities in higher education. As students and administrators turn their attention to science, technology, engineering, and mathematics, the humanities have struggled to define and defend their relevance, and the environmental humanities gives them an institutional and intellectual framework for doing so.

Although the concept of environment is undoubtedly central to the environmental humanities, there is something of a paradox in the field's engagement with the concept. Some of its most prominent scholars either seem uninterested in the concept or, when they do pause to reflect critically on it, conclude that it is of dubious value. For the former group, the term *environment* suffices as a convenient label for a wide range of concerns about the material conditions of life on Earth, particularly when those conditions are being changed for the worse by human actions. In other words, it conveys an interest in the kinds of issues conventionally associated with the environmental movement. It is a label that is readily legible to colleagues, students, administrators, and grantmaking agencies, they conclude, and to the extent that it suffices to attract resources and bring people with common interests into conversation, further critical reflection seems unnecessary and perhaps even counterproductive. As Rob Nixon has argued of the term *environmental humanities*, the most important consideration is "what kind of work is it enabling, rather than trying to finesse from an intellectual stance whether it's the correct term or the incorrect term or whether it's superannuated."[48] Even those scholars who believe that the concept of environment is both important and complex enough to warrant careful study often arrive at the same conclusion. Generally focusing on only the most recent and dominant variations, they conclude—quite reasonably, given the limits they have set on their inquiry—that "environment" is a poor fit to today's circumstances and aims. Nonetheless, even as they propose various experimental alternatives, most of them continue to describe their work as "environmental," for much the same practical reasons that the first group does.

The curious result is that most scholars in the environmental humanities seem to have tacitly agreed that "environment" is a convenient label, and little more.

By situating the environmental humanities in relation to the diverse forms of environmentalism described in the previous chapters, we can see how the critical impulse that has driven some scholars to challenge the concept of environment might be turned in a more productive direction. Instead of critiquing a variation on the concept of environment and an associated form of environmentalism that they have defined in advance—usually based on the "official" environmentalism of government agencies and NGOs—scholars in the environmental humanities might deploy their critical tools to better understand the diverse environments and environmentalisms that actually exist in the world and to advocate for those that seem most promising. A few scholars have already begun explicitly doing such work, while many others are doing it implicitly. Embracing environmental pluralism frees us from the need to decide once and for all on the "best" way of thinking and acting environmentally—a task as likely to be as unhelpful as it is difficult. Rather, because scholars in the humanities are trained to question fundamental concepts and to be sensitive to subtle nuances in meaning, they are uniquely well equipped to serve as interpreters of and mediators between radically different ways of encountering, experiencing, and representing our environments. Whether the environmental humanities will be able to play that mediating role depends not only on choices that humanities scholars are making in their scholarship, teaching, and public engagement efforts, but also on whether the world is interested in hearing what they have to say.

*

The historiography of environmentalism is often practiced as the search for the origins of environmentalism either as it exists today or as the historian wishes it might be. For the most part, such histories assume that what an environment *is* has always been obvious; the only thing up for debate is how that self-evident entity has been imagined, managed, damaged, improved, or otherwise affected by humanity. To the extent that historians have sought to historicize the environment itself as an object of knowledge and concern, they have tended to locate its origins in the emerging environmental movement of the mid-twentieth century. This book has taken a different approach: it has sought to understand what people in the past have meant and how they have acted when they claimed to understand their surroundings as "an environment" (or "the environment"). It has argued, moreover, that by broadening the definition of *environmentalist* to include everyone who approaches the world in terms of environments, we can also learn something about environmentalism in the more conventional sense of the word. In particular, we can learn that the way of thinking about and acting toward one's surroundings that has conventionally been defined as "environmentalist" is only one of the many ways that people have sought to improve their relationships their surroundings, and that one of conventional environmentalism's biggest flaws may be its tendency to assume that it offers the only proper way of doing so.

At its most basic, then, this book is an attempt to demonstrate in historical terms the value of environmental pluralism—that is, the idea that an important step toward ensuring that our surroundings support the kinds of lives we want to lead is recognizing that our own way of encountering those surroundings is only one of many possible ways, each of which may be appropriate to particular circumstances and aims. To be clear, this is not to say that anything goes, or that any particular variant on the concept of environment is as good as any other. On the contrary, it makes our criteria for a good environmentalism even more stringent, since such an environmentalism must be precisely tailored to meet the circumstances and aims of a particular time and place as well as comprehensible to people in other times and places. Moreover, given that even in our diversity there are many things we still share and that many of the circumstances that shaped past environmentalisms continue to operate today, there are some desiderata that any emerging environmentalism should satisfy, be it the scientific environmentalism of the critical zone, the indigenous environmentalism of the #NoDAPL movement, the artistic environmentalism of the Aerocene, or the scholarly environmentalism of the environmental humanities. We will, for example, probably want such environmentalisms to acknowledge that the boundaries of the entities they are concerned with are permeable and changeable, to reckon with the ongoing legacies of European colonialism and the forms of scientific racism and social hierarchy associated with them, to provide a convincing account of the entanglement of social and physical environments, and to be capable of moving between multiple scales of time and space.

Finally, even as we look for variants on the concept of environment and the environmentalisms that correspond to them, we will want to find environmentalisms that not only recognize their own limits but also seek to form productive and ethical connections to other environmentalisms. It is here that scholarship in the environmental humanities in general and in the discipline of history in particular may be of some use. If the history of past environments and environmentalisms reveals anything, it is that no environmentalism is sufficient unto itself; every form of environmentalism that has sought to become universal has inevitably done violence both to living beings and to the conditions those beings require to survive and to thrive. The more we know about the range of environmentalisms that have existed in the past—each materializing its own distinct variation on the concept of environment—the less likely it is that we will be seduced by the mirage of a single, perfect, all-encompassing environmentalism. In short, recalling the multiple pasts of environmentalism may be one of the most effective means we have of nurturing its many possible futures.

ACKNOWLEDGMENTS

Scholarship can only thrive in a supportive environment, and the University of Pennsylvania's Department of History and Sociology of Science has been a very supportive environment for this particular scholarly project. I am grateful to my current and former colleagues Robert Aronowitz, David Barnes, Meggie Crnic, Stephanie Dick, Sebastián Gil-Riaño, Ann Norton Greene, Andi Johnson, Robert Kohler, Harun Kuçuk, Susan Lindee, Beth Linker, Ramah McKay, Jonathan Moreno, Projit Mukharji, John Tresch, and Adelheid Voskuhl. This book would not be what it is without everything you have taught me.

I think my colleagues would agree that, however excellent we may be as scholars and teachers, our department owes much of its spark to our graduate students. I feel lucky to have been able to work with and learn from Cameron Brinitzer, Jason Chernesky, Sumiko Hatakeyama, Sara Meloni, Mary X. Mitchell, Anna Mucser, Jeffrey Nagle, Lisa Ruth Rand, Sara Ray, Alexis Rider, Jesse Smith, and Nicole Welk-Joerger, as well as doctoral students elsewhere at Penn and beyond, including Dalal Musaed Alsayer, Hannah Anderson, Emily Leifer, Aylin Malcolm, Pooja Nayak, and Martin Premoli. Thanks to all of you for inviting me into your scholarship.

It is easy to experience teaching as a burden when one is trying and sometimes failing to write a book, but it really does make scholarship better, particularly for a big-picture project of this sort. I am grateful to all my students, especially to the undergraduates in Environment and Society (STSC-168), where many of the ideas in this book were first articulated. I am sure that you did not expect to spend so much of the semester parsing fine distinctions between "nature" and "environment" or learning about eighteenth-century natural history. Thank you for bearing with me and helping me figure out why it matters.

Slow in the making, this book would have been even slower without several much-needed injections of research funding and teaching relief. As chair of the Department of History and Sociology of Science, Robert Aronowitz helped free up time for writing at a critical juncture. Thank you for your moral and material support. This project also benefited from a Research Opportunity Grant from Penn's School of Arts and Sciences, from research funds associated with the Janice and Julian Bers Assistant Professorship in the Social Sciences, and from faculty fellowships from the Price Lab for Digital Humanities and the Wolf Humanities Center.

I owe a special debt of gratitude to Bethany Wiggin, founding director of the Penn Program in Environmental Humanities (PPEH), who joined me in organizing what turned out to be an extraordinarily stimulating faculty working group from 2016 to 2018, many of whose discussions shaped the arguments offered in this book, and who invited me to serve as PPEH topic director for a year of presentations and conversations around the question "How Did We Get Here?" Thank you for creating a space where urgent, interdisciplinary, public-facing, arts-oriented scholarship can flourish.

Other institutions beyond Penn were also important to the development of this project. The idea of a critical history of the concept of environment first occurred to me during the time I spent as a research fellow in Department II of the Max Planck Institute for the History of Science in Berlin before departing for Philadelphia. Lorraine Daston, the director of that department from 1995 to 2019, has shown a generation of historians of science not only how to historicize such taken-for-granted concepts as probability, wonder, objectivity, observation, data, and rationality but also how to do so with rigor, style, and good humor. Thank you for raising the bar.

Although I only spent three months at the Rachel Carson Center for Environment and Society in Munich in the fall of 2014, I am also grateful to that institution and to its directors, Christof Mauch and Helmut Trischler, who in a few short years managed *ein weltberühmtes Forschungszentrum aus dem Boden zu stampfen*, as they say. I workshopped an early proposal for this project during my short visit there amid an extraordinary group of international scholars, some of whom have become lasting friends.

It is a truism that the seemingly solitary act of writing a book is in fact a collaboration among many scholars past and present, but that claim is perhaps especially true in regard to a synthetic project such as this one. In particular,

I have learned a tremendous amount—sometimes up close, sometimes at a distance—from the work of scholars in the borderlands between the history and philosophy of science, environmental history, and science and technology studies, including Peter Alagona, Mark Barrow, Angela Creager, Raf de Bont, Stefan Helmreich, Sheila Jasanoff, Robert Kohler, Gregg Mitman, Michelle Murphy, Linda Nash, Trevor Pearce, Harriet Ritvo, Christopher Sellers, Sverker Sörlin, Jeremy Vetter, Nina Wormbs, and Donald Worster. Thank you for showing that it is possible to care deeply about the environment even while demonstrating how forms of care and understandings of the environment are contingent and contestable.

As I was putting the final touches on this book, two others on closely related topics appeared that supported some of my conclusions, forced me to revise others, and reminded me that the history of the environment is far too big to be encompassed by any one book. Those two books are Paul Warde, Libby Robin, and Sverker Sörlin's *The Environment: A History of the Idea*, some of whose key claims I had encountered in a series of earlier articles and in their edited volume *The Future of Nature*, and Perrin Selcer's *The Postwar Origins of the Global Environment*. Other scholarly debts will be evident throughout the book. Thank you for starting such a rich, inviting, and important scholarly conversation.

As should have been obvious, but as I only belatedly discovered in the early stages of this project, it is possible to share a project too soon. My thanks and apologies to the audiences who heard early versions at the Massachusetts Institute of Technology, the Center for 21st Century Studies at the University of Wisconsin–Milwaukee, the Max Planck Institute for the History of Science, Johns Hopkins University, and the architecture PhD students' colloquium at the University of Pennsylvania. While I cannot give you back those hours of your life, I hope there is some small consolation (very small, I'm sure) in knowing that your hard questions and skeptical looks made this a better book. I am also grateful to Lukas Rieppel and the other organizers of and participants in the Political Concepts: The Science Edition conference at Brown University, who helped me clarify key claims as this project was nearing completion.

It is customary for historians to acknowledge the hardworking archivists who make our scholarship possible, and since they cannot be thanked enough, I will gladly take the opportunity to do so once again. Nonetheless, this was not the sort of project that required a deep dive into the archive. Instead it depended on truckloads of books, newspapers, reports, and other published materials. Not surprisingly given the age we live in, most of these truckloads were virtual. Among the many sources of digitized texts that have become available in the past several decades, I am especially grateful to the Internet Archive—perhaps one of the few corners of the internet that retains a glimmer of the medium's early promise. For the physical truckloads, I am grateful to the Penn library system, particularly the indefatigable interlibrary loan staff. Your work is why libraries remain spaces of utopian possibility.

It is not an overstatement to say that I could not have finished this project without the assistance of Alexis Rider, who over the course of a year helped me transform footnotes ranging from "insert source here"-style placeholders to multipage "see also" bibliographies into useful citations, sometimes in ways that led me to fundamentally revise my arguments. She did so, moreover, while developing a fascinating dissertation project on the history of the science of ice. I can't wait to read it. I am also grateful to Peter Collopy and Ivan Sandoval, who at an early stage of the project helped me consider possible avenues for research. Even though those were not the paths I ended up following, they shaped the form the project eventually took. Thank you for going exploring with me.

A number of people read and commented on most or all of the manuscript in one or several of its iterations, including Robert Aronowitz, Cameron Brinitzer, Bernard Dionysius Geoghegan, John Tresch, and the anonymous reviewers for the University of Chicago Press. Ekaterina Babintseva generously read the chapter on the biosphere and shared her insights into an important passage from Vernadskii. At key moments, conversations with Fredrik Albritton Jonsson, Mary X. Mitchell, Nicholas Shapiro, and John Tresch ranging in length from a few minutes to a few hours helped me clarify the stakes. I may not have seemed grateful for your feedback in the moment, but I was and am. Your enthusiasm helped me keep going with the project, and your probing questions and gentle nudges helped me change direction when I needed to.

It has been a pleasure to work with Karen Merikangas Darling at the University of Chicago Press to bring this book to print. Thank you for taking a chance on what must have at the beginning seemed like a hopelessly overambitious proposal, and for insisting that I could say what I need to say without inventing ugly new words to say it—even if this book is, indeed, about processes of environing that bring new kinds of environees and environments into being. I am also grateful to the rest of the University of Chicago Press team, as well as to Johanna Rosenbohm for copyediting and Derek Gottlieb for indexing. Any remaining errors or infelicities are mine and mine alone.

Above all, I want to thank Tina, my companion at every fork in the road, and Frida Claire, who has been surprising and delighting us since the day she was born, two weeks after the final draft of the manuscript was due and many months before it was finished. Thanks to both of you for reminding me that books are wonderful things and that books are not everything. This one is dedicated to you.

NOTES

INTRODUCTION

1. Claude Bernard, *Leçons sur les phénomènes de la vie communes aux animaux et aux végétaux* (Paris: J.-B. Baillière et Fils, 1879), 4–5. See also Frederic L. Holmes, "Claude Bernard, the 'Milieu Intérieur,' and Regulatory Physiology," *History and Philosophy of the Life Sciences* 8, no. 1 (1986): 3–25.

2. Starbucks, "Starbucks Announces Global Greener Stores Commitment," press release, September 13, 2018, https://stories.starbucks.com/press/2018/starbucks-announces-global -greener-stores-commitment/.

3. Jenny Price, "Remaking American Environmentalism: On the Banks of the L.A. River," *Environmental History* 13, no. 3 (2008): 536–55, quote on 540.

4. See, e.g., William Cronon, "The Trouble with Wilderness: Or, Getting Back to the Wrong Nature," *Environmental History* 1, no. 1 (1996): 7–28.

5. Lewis Herber [Murray Bookchin], *Our Synthetic Environment* (New York: Knopf, 1962).

6. Murray Bookchin, "Toward an Ecological Society," in *Toward an Ecological Society* (Montreal: Black Rose Books, 1980), 56–71, on 58.

7. Richard Lewontin, "Gene, Organism, and Environment," in *Cycles of Contingency: Developmental Systems and Evolution*, ed. Susan Oyama, Paul E. Griffiths, and Russell D. Gray (Cambridge, MA: MIT Press, 2001), 1–11; first published in *Evolution: From Molecules to Men*, ed. D. S. Bendall (Cambridge: Cambridge University Press, 1983), 273–85. See also Richard C. Lewontin, *The Triple Helix: Gene, Organism, and Environment* (Cambridge, MA: Harvard University Press, 2000).

8. Georges Canguilhem, "The Living and Its Milieu," in *Knowledge of Life*, ed. Paola Marrati and Todd Meyers, trans. Stefanos Geroulanos and Daniela Ginsburg (New York: Fordham University Press, 2008), 98–120; Tim Ingold, "Globes and Spheres: The Topology of Environmentalism," in *Environmentalism: The View from Anthropology* (New York: Routledge, 1993),

31–42; Linda L. Nash, *Inescapable Ecologies: A History of Environment, Disease, and Knowledge* (Berkeley: University of California Press, 2006). See also Steven Kroll-Smith and Worth Lancaster, "Review: Bodies, Environments, and a New Style of Reasoning," *Annals of the American Academy of Political and Social Science* 584 (2002): 203–12.

9. Ingold, "Globes and Spheres," 40.

10. See, e.g., Bruno Latour, *Facing Gaia: Eight Lectures on the New Climatic Regime*, trans. Catherine Porter (Medford, MA: Polity, 2017); Donna J. Haraway, *Staying with the Trouble: Making Kin in the Chthulucene* (Durham, NC: Duke University Press, 2016); Stacy Alaimo, *Bodily Natures: Science, Environment, and the Material Self* (Bloomington: Indiana University Press, 2010); Timothy Morton, *Ecology without Nature: Rethinking Environmental Aesthetics* (Cambridge, MA: Harvard University Press, 2007); Timothy Morton, *The Ecological Thought* (Cambridge, MA: Harvard University Press, 2010); and Jane Bennett, *Vibrant Matter: A Political Ecology of Things* (Durham, NC: Duke University Press, 2010).

11. Bruno Latour, "Why Gaia Is Not a God of Totality," *Theory, Culture & Society* 34, no. 2–3 (2017): 61–81, quote on 74.

12. Haraway, *Staying with the Trouble*, 30.

13. For early examples, see Roderick Nash, *The American Environment: Readings in the History of Conservation* (Reading, MA: Addison-Wesley, 1968); Donald Worster, ed., *American Environmentalism: The Formative Period, 1860–1915* (New York: Wiley, 1973); and Samuel P. Hays in collaboration with Barbara D. Hays, *Beauty, Health, and Permanence: Environmental Politics in the United States, 1955–1985* (New York: Cambridge University Press, 1987).

14. See, e.g., Joan W. Scott, "Gender: A Useful Category of Historical Analysis," *American Historical Review* 91, no. 5. (1986): 1053–75; Leo Marx, "'Technology': The Emergence of a Hazardous Concept," *Social Research* 64, no. 3 (1997): 965–88; and James Clifford and George E. Marcus, eds., *Writing Culture: The Poetics and Politics of Ethnography* (Berkeley: University of California Press, 1986).

15. William Cronon, ed., *Uncommon Ground: Toward Reinventing Nature* (New York: W. W. Norton, 1995). See especially William Cronon, "The Trouble with Wilderness: Or, Getting Back to the Wrong Nature," ibid., 69–90.

16. Paul S. Sutter, "The World with Us: The State of American Environmental History," *Journal of American History* 100, no. 1 (2013): 94–119.

17. See, e.g., Richard H. Grove, *Green Imperialism: Colonial Expansion, Tropical Island Edens, and the Origins of Environmentalism, 1600–1860* (New York: Cambridge University Press, 1995); Aaron Sachs, *The Humboldt Current: Nineteenth-Century Exploration and the Roots of American Environmentalism* (New York: Viking, 2006); Harriet Ritvo, *The Dawn of Green: Manchester, Thirlmere, and Modern Environmentalism* (Chicago: University of Chicago Press, 2009); and Fredrik Albritton Jonsson, *Enlightenment's Frontier: The Scottish Highlands and the Origins of Environmentalism* (New Haven, CT: Yale University Press, 2013).

18. Key works in this emerging literature include Michelle Murphy, *Sick Building Syndrome and the Problem of Uncertainty: Environmental Politics, Technoscience, and Women Workers* (Durham, NC: Duke University Press, 2006); Christopher C. Sellers, *Crabgrass Crucible: Suburban Nature and the Rise of Environmentalism in Twentieth-Century America* (Chapel Hill: University of North Carolina Press, 2012); Joshua P. Howe, *Behind the Curve: Science and the*

Politics of Global Warming (Seattle: University of Washington Press, 2014); Christophe Bonneuil and Jean-Baptiste Fressoz, *The Shock of the Anthropocene: The Earth, History and Us*, trans. David Fernbach (New York: Verso, 2015); Paul Warde, Libby Robin, and Sverker Sörlin, *The Environment: A History of the Idea* (Baltimore, MD: Johns Hopkins University Press, 2018); and Perrin Selcer, *The Postwar Origins of the Global Environment: How the United Nations Built Spaceship Earth* (New York: Columbia University Press, 2018).

19. See Libby Robin, Sverker Sörlin, and Paul Warde, eds., *The Future of Nature: Documents of Global Change* (New Haven, CT: Yale University Press, 2013); Warde, Robin, and Sörlin, *Environment*; Sverker Sörlin, "Reconfiguring Environmental Expertise," *Environmental Science & Policy* 28 (2013): 14–24; Paul Warde, Libby Robin, and Sverker Sörlin, "Stratigraphy for the Renaissance: Questions of Expertise for 'the Environment' and 'the Anthropocene,'" *Anthropocene Review* 4, no. 3 (2017): 246–58; Paul Warde, "The Environmental History of Preindustrial Agriculture in Europe," in *Nature's End: History and the Environment*, ed. Sverker Sörlin and Paul Warde (New York: Palgrave Macmillan, 2009), 70–92; and Sverker Sörlin and Nina Wormbs, "Environing Technologies: A Theory of Making Environment," *History and Technology* 34, no. 2 (2018): 101–25.

20. *Oxford English Dictionary*, online ed., s.v. "environment," accessed August 31, 2019.

21. Thomas Carlyle, *The Life of John Sterling* (London: Chapman and Hall, 1851), 146. See also Ralph Jessop, "Coinage of the Term Environment: A Word without Authority and Carlyle's Displacement of the Mechanical Metaphor," *Literature Compass* 9, no. 11 (2012): 708–20.

22. *Oxford English Dictionary*, s.v. "environment."

23. See, e.g., Arthur O. Lovejoy, *The Great Chain of Being: A Study of the History of an Idea* (Cambridge, MA: Harvard University Press, 1936); and Reinhart Koselleck, *The Practice of Conceptual History: Timing History, Spacing Concepts*, trans. Todd Samuel Presner et al. (Stanford, CA: Stanford University Press, 2002).

24. Harriet Martineau, *The Positive Philosophy of Auguste Comte* (London: J. Chapman, 1853). See also Trevor Pearce, "From 'Circumstances' to 'Environment': Herbert Spencer and the Origins of the Idea of Organism-Environment Interaction," *Studies in History and Philosophy of Biological and Biomedical Sciences* 41, no. 3 (2010): 241–52; and Trevor Pearce, "The Origins and Development of the Idea of Organism-Environment Interaction," in *Entangled Life: Organism and Environment in the Biological and Social Sciences*, ed. Gillian Barker, Eric Desjardins, and Trevor Pearce (Dordrecht: Springer, 2014), 13–32.

25. John Tresch, "Cosmologies Materialized: History of Science and History of Ideas," in *Rethinking Modern European Intellectual History*, ed. Darrin McMahon and Sam Moyn (New York: Oxford University Press, 2014), 153–72. The language of "materialization" also figures prominently in Murphy, *Sick Building Syndrome*. A similar approach, albeit using a different idiom, is pursued by Jennifer Gabrys in her study of the "becoming environmental" of computing; see Jennifer Gabrys, *Program Earth: Environmental Sensing Technology and the Making of a Computational Planet* (Minneapolis: University of Minnesota Press, 2016), 9.

26. Lynton K. Caldwell, "Environment: A New Focus for Public Policy?," *Public Administration Review* 23, no. 3 (1963): 132–39, quote on 133.

27. Michel Foucault, *Security, Territory, Population: Lectures at the Collège de France, 1977–78*, ed. Michel Senellart, trans. Graham Burchell (New York: Palgrave Macmillan, 2007), 21.

CHAPTER 1

1. *Lorient* is sometimes written as *l'Orient*. According to the French Republican calendar, the *Géographe* arrived on the third day of the month of Floréal in year 12; Étienne Geoffroy Saint-Hilaire, "Note sur les animaux vivans venus à bord du Géographe," *Annales du Muséum National d'Histoire Naturelle* 4 (1804): 171–72, on 171.

2. A. L. Jussieu, "Notice sur l'expédition à la Nouvelle-Hollande, Entreprise pour des recherches de Géographie et d'Histoire naturelle," *Annales du Muséum National d'Histoire Naturelle* 5 (1804): 1–11, on 3.

3. Nicole Starbuck, *Baudin, Napoleon, and the Exploration of Australia* (London: Pickering & Chatto, 2013), 137.

4. The *Naturaliste* arrived in France eight months before the *Géographe*; Geoffroy, "Note sur les animaux vivans," 171–72.

5. Joseph Philippe François Deleuze, *Histoire et description du Muséum royal d'histoire naturelle* (Paris: A. Royer, 1823), 105. See also Starbuck, *Baudin, Napoleon, and Australia*, 1.

6. Geoffroy, "Note sur les animaux vivans," 171–72. See also Starbuck, *Baudin, Napoleon, and Australia*, 133.

7. Jussieu, "Notice sur l'expédition à la Nouvelle-Hollande," 7.

8. Deleuze, *Histoire et description du Muséum*, 105–8.

9. Deleuze, 87–88.

10. Nicholas Jardine, James A. Secord, and Emma C. Spary, eds., *Cultures of Natural History* (New York: Cambridge University Press, 1996); Helen A. Curry, Nicholas Jardine, James A. Secord, and Emma C. Spary, eds., *Worlds of Natural History* (New York: Cambridge University Press, 2018).

11. A. L. Jussieu, "Notice Historique sur le Muséum d'Histoire Naturelle," *Annales du Muséum National d'Histoire Naturelle* 1 (1802): 1–14. See also Pierre-Yves Lacour, introduction to *La République naturaliste: Collections d'histoire naturelle et Révolution française (1789-1804)* (Paris: Publications Scientifique du Muséum national d'Histoire naturelle, 2014), 7–40, esp. 19.

12. Jacques Roger, *Buffon: A Life in Natural History*, ed. L. Pearce Williams, trans. Sarah Lucille Bonnefoi (Ithaca, NY: Cornell University Press, 1997).

13. See Louise E. Robbins, *Elephant Slaves and Pampered Parrots: Exotic Animals in Eighteenth-Century Paris* (Baltimore: Johns Hopkins University Press, 2002); and Anita Guerrini, *The Courtiers' Anatomists: Animals and Humans in Louis XIV's Paris* (Chicago: University of Chicago Press, 2015).

14. E. C. Spary, *Utopia's Garden: French Natural History from Old Regime to Revolution* (Chicago: University of Chicago Press, 2000), 193–240.

15. B. G. E. L. Lacépède, G. Cuvier, and J.-B. Lamarck, "Rapport des Professeurs du Muséum sur les Collections d'Histoire Naturelle Rapportées d'Égypte, par E. Geoffroy," *Annales du Muséum National d'Histoire Naturelle* 1 (1802): 234–41. See also Tony Appel, *The Cuvier-Geoffroy Debate: French Biology in the Decades before Darwin* (Oxford: Oxford University Press, 1987), 72–74; and Dorinda Outram, *Georges Cuvier: Vocation, Science, and Authority in Post-Revolutionary France* (Manchester: Manchester University Press, 1984), 61–62.

16. Florence F. J. M Pieters, "Notes on the Menagerie and Zoological Cabinet of Stadholder

William V of Holland, directed by Aernout Vosmaer," *Journal of the Society for the Bibliography of Natural History* 9, no. 4 (1980): 539–63. See also Outram, *Georges Cuvier*, 164–65.

17. Lacour, *La République naturaliste*, 219.

18. André Thouin to unknown correspondent, 4 ventôse an 3 [February 22, 1795], F17A/1276/dossier 2, Archives Nationales, Paris, as quoted and translated in Elise Lipkowitz, "The 'Elephant in the Room': The Impact of the French Seizure of the Dutch Stadholder's Collection on Relations between Dutch and French Naturalists," in *Of Elephants & Roses: French Natural History, 1790–1830*, ed. Sue Ann Prince (Philadelphia: American Philosophical Society, 2013), 101–9, on 104.

19. Richard W. Burkhardt Jr., *The Spirit of System: Lamarck and Evolutionary Biology* (Cambridge, MA: Harvard University Press, 1995), 119.

20. Muséum Royal d'Histoire Naturelle, *Instructions pour les Voyageurs et pour les Employés dans les Colonies, sur la Manière de Recueillir, de Conserver et d'Envoyer les Objets d'Histoire Naturelle* (A. Belin: Paris, 1818), 1.

21. Muséum Royal d'Histoire Naturelle, *Instructions pour les Voyageurs*, 6.

22. Lee Alan Dugatkin, *Mr. Jefferson and the Giant Moose: Natural History in Early America* (Chicago: University of Chicago Press, 2009), 99.

23. Thomas Jefferson, "Description d'une oreille de charrue, offrant le moins de résistance possible, et dont l'exécution est aussi facile que certain," *Annales du Muséum National d'Histoire Naturelle* 1 (1802): 322–31; Thomas Jefferson, "Extrait d'une lettre de M. Thomas Jefferson, président des Etats-Unis de l'Amerique," *Annales du Muséum National d'Histoire Naturelle* 5 (1804): 316.

24. Dugatkin, *Mr. Jefferson and the Giant Moose*, 59–61; Joseph J. Ellis, *American Sphinx: The Character of Thomas Jefferson* (New York: Alfred A. Knopf, 1997), 242–52.

25. Starbuck, *Baudin, Napoleon, and Australia*, 103.

26. François Péron and Louis Freycinet, *Voyage de découvertes aux terres Australes, exécuté sur les corvettes le Géographe, le Naturaliste, et la goélette le Casuarina, pendant les années 1800, 1801, 1802, 1803 et 1804*, vol. 2 (Paris: L'Imprimerie Royale, 1816), 35.

27. François Péron, *Voyage de découvertes aux terres Australes, exécuté par ordre de Sa Majesté l'Empereur et Roi, sur les Corvettes le Géographe, le Naturaliste, et la Goelette le Casuarina pendant les Années 1800, 1801, 1802, 1803 et 1804*, vol. 1 (Paris: l'Imprimerie Impériale, 1807), 154–55. See also Starbuck, *Baudin, Napoleon, and Australia*, 111.

28. Péron, *Voyage de découvertes aux terres Australes*, vol. 1, 445–84.

29. Richard W. Burkhardt Jr., "Unpacking Baudin: Models of Scientific Practice in the Age of Lamarck," in *Jean-Baptiste Lamarck, 1744–1829*, ed. Goulven Laurent (Paris: Editions du CTHS, 1997), 497–514.

30. Georges-Louis Leclerc, Comte de Buffon, "Premier discours: De la manière d'étudier & de traiter l'Histoire Naturelle," in *Histoire naturelle, générale et particulière, avec la description du Cabinet du Roi*, vol. 1 (Paris: Imprimerie Royale, 1749), 3–65, esp. 11–13.

31. E.-T. Hamy, "Les derniers jours du Jardin du Roi et la fondation du Muséum d'Histoire Naturelle," in *Centenaire de la Fondation du Muséum d'Histoire Naturelle, 10 juin 1793 – 10 juin 1893* (Paris: Imprimerie National, 1893), 1–162, on 4; Deleuze, *Histoire et description du Muséum*, 83.

32. Richard W. Burkhardt Jr., "Civilizing Specimens and Citizens at the Muséum d'Histoire Naturelle, 1793–1838," in *Of Elephants & Roses*, ed. Prince, 14–30, esp. 19–20.

33. Quoted in Burkhardt, "Civilizing Specimens and Citizens," 19.

34. On the acquisition of the hackney-carriage administration lot, which had subsequently been used for a flour warehouse, see Deleuze, *Histoire et description du Muséum*, 85.

35. On the expansion of the galleries, see Deleuze, *Histoire et description du Muséum*, 127.

36. Georges Cuvier, "Notice sur l'établissement de la collection d'anatomie comparée du Muséum," *Annales du Muséum National d'Histoire Naturelle* 2 (1803): 409–14, quote on 411–12.

37. Deleuze, *Histoire et description du Muséum*, 127.

38. See, e.g., Henri de Blainville, "Mémoire sur le squale pèlerin," *Annales du Muséum d'Histoire Naturelle* 18 (1811): 88–134.

39. André Thouin, "Description du Jardin des Semis du Muséum d'Histoire naturelle, de sa culture et de ses usages, première partie," *Annales du Muséum National d'Histoire Naturelle* 4 (1804): 263–88.

40. Robbins, *Elephant Slaves and Pampered Parrots*; Michael A. Osborne, "Zoos in the Family: The Geoffroy Saint-Hilaire Clan and the Three Zoos of Paris," in *New Worlds, New Animals: From the Menagerie to Zoological Park in the Nineteenth Century*, ed. R. J. Hoage and William A. Deiss (Baltimore: Johns Hopkins University Press, 1996), 33–42.

41. Richard W. Burkhardt Jr., "La Ménagerie et la vie du Muséum," in *Le Muséum au Premier Siècle de Son Histoire*, ed. Claude Blanckaert (Paris: Editions du Muséum National d'Histoire Naturelle, 1997), 481–508; Outram, *Georges Cuvier*, 57.

42. Richard W. Burkhardt Jr., "The Leopard in the Garden: Life in Close Quarters at the Muséum d'Histoire Naturelle," *Isis* 98, no. 4 (2007): 675–94.

43. Georges Cuvier, "Recherches sur les espèces vivantes de grands chats, pour servir de preuves et d'éclaircissements au chapitre sur les Carnassiers fossiles," *Annales du Muséum d'Histoire Naturelle* 14 (1809): 136–64, on 152–53. See also Appel, *Cuvier-Geoffroy Debate*, 36.

44. Pierre-Yves Lacour, "Picturing Nature in a Natural History Museum: The Engravings of the *Annales du Muséum d'Histoire Naturelle*, 1802–13," in *Of Elephants & Roses*, ed. Prince, 117–27. See also Daniela Bleichmar, *Visible Empire: Botanical Expeditions and Visual Culture in the Hispanic Enlightenment* (Chicago: University of Chicago Press, 2012), 154.

45. Christopher M. Parsons and Kathleen S. Murphy, "Ecosystems under Sail: Specimen Transport in the Eighteenth-Century French and British Atlantics," *Early American Studies* 10, no. 3 (2012): 503–29.

46. Jussieu, "Notice sur l'Expédition à la Nouvelle-Hollande," 4.

47. Starbuck, *Baudin, Napoleon, and Australia*, 133.

48. Jussieu, "Notice sur l'Expédition à la Nouvelle-Hollande," 6–7.

49. Thouin, "Description du Jardin des Semis, première partie," 265, 287–88.

50. André Thouin, "Description du Jardin des semis du Muséum, seconde partie," *Annales du Muséum d'Histoire Naturelle* 6 (1805): 172–96, quote on 196.

51. Thouin, "Description du Jardin des Semis, première partie," 274–80, quote on 274–75.

52. A. Thouin, "Description et usage de plusieurs ustensiles de moderne invention, propres

à la culture d'un grand nombre de plantes dans les écoles de botaniques," *Annales du Muséum d'Histoire Naturelle* 6 (1805): 236–52, on 236.

53. Thouin, "Description du Jardin des Semis, première partie," 270–71.

54. A. Thouin, "Description et usage de plusieurs ustensiles de moderne invention," 238. Greenhouses were also used to grow rare plants; see R. L. Desfontaines, "Plantes rares qui ont fleuri en L'an X dans le jardin ou dans les serres du Muséum," *Annales du Muséum d'Histoire Naturelle* 1 (1802): 127–34, 200–206, 276–80.

55. Thouin, "Description et usage de plusieurs ustensiles," 249–52.

56. J. P. F. Deleuze, "Suite du memoire sur les plantes d'ornement, et sur leur introduction dans nos jardins," *Annales du Muséum d'Histoire Naturelle* 9 (1807): 149–204, on 200–204.

57. Deleuze, "Suite du memoire sur les plantes d'ornement," 201–2.

58. "Extrait d'une lettre, en date du 4 septembre 1808, écrite par M. Joseph Martin, directeur des cultures coloniales des arbres à épiceries à Cayenne, et correspondant du Muséum d'Histoire naturelle, au professeur Thouin," *Annales du Muséum d'Histoire Naturelle* 12 (1808), 460–63.

59. E.g., Frédéric Cuvier, "Observations sur le chien des habitans de la Nouvelle-Hollande, précédées de quelques réflexions sur les facultés morales des animaux," *Annales du Muséum d'Histoire Naturelle* 11 (1808): 458–76.

60. Cuvier, "Observations sur le chien," 475; Frédéric Cuvier, "Note sur l'accouplement d'un zèbre et d'un cheval," *Annales du Muséum National d'Histoire Naturelle* 11 (1808), 237–40; Frédéric Cuvier, "Observations sur l'accouplement d'un Cygne chanteur mâle et d'une Oie domestique femelle, et Description du Mulet qui en est provenu," *Annales du Muséum d'Histoire Naturelle* 12 (1808), 119–25; Étienne Geoffroy Saint-Hilaire, "Mouvemens de la Ménagerie," *Annales du Muséum National d'Histoire Naturelle* 4 (1804), 94–104, on 102–3.

61. Appel, *Cuvier-Geoffroy Debate*, 75–77.

62. Tobias Cheung, "What Is an 'Organism'? On the Occurrence of a New Term and Its Conceptual Transformations 1680–1850," *History and Philosophy of the Life Sciences* 32, no. 2/3 (2010): 155–94, on 184; Georges Canguilhem, "The Living and Its Milieu," in *Knowledge of Life*, ed. Paola Marrati and Todd Meyers, trans. Stefanos Geroulanos and Daniela Ginsburg (New York: Fordham University Press, 2008), 98–120.

63. Jean-Baptiste Lamarck, *Recherches sur l'organisation des corps vivans, et particulairèment sur son origine, sur la cause de ses développmens et des progrès de sa composition, et sur celle qui, tendant continuellement à la détruire dans chaque individu, amène nécessairement sa mort; Précédé du Discours d'ouverture du cours de Zoologie, donné dans le muséum national d'Histoire Naturelle, l'an X de la république* (Paris: Maillard, 1802), 202. See also Giulio Barsanti, "Lamarck et la naissance de la Biologie," in *Jean-Baptiste Lamarck, 1744–1829*, ed. Laurent, 349–68; and Joan Steigerwald, "Treviranus' *Biology*: Generation, Degeneration, and the Boundaries of Life," in *Reproduction, Race, and Gender in Philosophy and the Early Life Sciences*, ed. Susanne Lettow (Albany: State University of New York Press, 2014), 105–27.

64. Jessica Riskin, *Restless Clock: A History of the Centuries-Long Argument over What Makes Living Things Tick* (Chicago: University of Chicago Press, 2016), 165.

65. Roger, *Buffon: A Life in Natural History*, 129.

66. Roger, *Buffon: A Life in Natural History*, 127–31; James L. Larson, *Interpreting Nature: The Science of Living Form from Linnaeus to Kant* (Baltimore: Johns Hopkins University Press, 1994), 139–37.

67. Roger, *Buffon: A Life in Natural History*, 299–301.

68. Georges-Louis Leclerc, Comte de Buffon, "De la dégénération des animaux," in *Histoire naturelle, générale et particulaire*, vol. 14 (Paris: Imprimerie Royale, 1766), 311–74, on 317. See also Roger, *Buffon: A Life in Natural History*, 300.

69. Peter J. Bowler, "Bonnet and Buffon: Theories of Generation and the Problem of Species," *Journal of the History of Biology* 6, no. 2 (1973): 259–81, on 269; Roger, *Buffon: A Life in Natural History*, 297.

70. Roger, *Buffon: A Life in Natural History*, 301.

71. Buffon, "Premier discours," 12–13. See also Arthur O. Lovejoy, *The Great Chain of Being: A Study of the History of an Idea* (Cambridge, MA: Harvard University Press, 1936).

72. Buffon, "Premier discours," 11. See also Roger, *Buffon: A Life in Natural History*, 292.

73. Burkhardt, *Spirit of System*; Pietro Corsi, *The Age of Lamarck: Evolutionary Theories in France, 1790–1830* (Berkeley: University of California Press, 1988); Appel, *Cuvier–Geoffroy Debate*; Outram, *Georges Cuvier*.

74. Lamarck, *Philosophie zoologique*, 380–81.

75. Georges Cuvier, *Le règne animal distribué d'après son organisation*, vol. 1 (Paris: A. Belin, 1817), 13.

76. Georges Cuvier, "Discours préliminaire," in *Recherches sur les ossemens fossiles de quadrupèdes, où l'on rétablit les caractères de plusieurs espèces d'animaux que les révolutions du globe paroissent avoir détruites*, vol. 1 (Paris: Déterville, 1812), 58.

77. Dorinda Outram, "Uncertain Legislator: Georges Cuvier's Laws of Nature in Their Intellectual Context," *Journal of the History of Biology* 19, no. 3 (1986): 323–68.

78. Cuvier, *Le règne animal*, 1:6.

79. Burkhardt, *Spirit of System*, 143–85.

80. Jean-Baptiste Lamarck, *Philosophie zoologique* (Paris: Muséum d'Histoire Naturelle, 1809), 223–25. See also Burkhardt, *Spirit of System*, 62.

81. Lamarck, *Philosophie zoologique*, 221–22. See also Canguilhem, "The Living and Its Milieu," 104.

82. Henri Marie Ducrotay de Blainville, *De l'organisation des animaux, où, principes d'anatomie comparée* (Paris: F. G. Levrault, 1822); Henri Marie Ducrotay de Blainville, *Cours de physiologie générale et comparée: Professé a la Faculté des Sciences de Paris*, 3 vols. (Paris: Germer Baillière, 1833). See also Bernard Balan, "Organisation, organisme, économie et milieu chez Henri Ducrotay de Blainville," *Revue d'histoire des sciences* 32, no. 1 (1979): 5–24.

83. *"Circonstances extérieures"* in the original; Blainville, *De l'organisation des animaux*, xlii–xliii.

84. Blainville, *Cours de physiologie générale et comparée*, vol. 1, 36–39. On Blainville's role in the development of the concepts of organism and milieu, see Tobias Cheung, *Organismen: Agenten zwischen Innen- und Außenwelten 1780–1860* (Bielefeld: Transcript Verlag, 2014), 238–48.

85. Auguste Comte, "Quarantième leçon: Considérations philosophiques sur l'ensemble de

la science biologique," in *Cours de philosophie positive*, vol. 3 (Paris: Bachelier, 1838), 269–486, on 301. Quoted (in slightly different translation) in Canguilhem, "The Living and Its Milieu," 101. See also Henri Gouhier, "Blainville et Auguste Comte," *Revue d'histoire des sciences* 32, no. 1 (1979): 59–72; Annie Petit, "L'héritage de Lamarck dans la philosophie positive d'Auguste Comte," in *Jean-Baptiste Lamarck*, ed. G. Laurent, 453–556; Jean-François Braunstein, "Le concept de milieu, de Lamarck à Comte et aux positivismes," in *Jean-Baptiste Lamarck*, ed. Laurent, 557–71; Chris McClellan, "The Legacy of Georges Cuvier in Auguste Comte's Natural Philosophy," *Studies in the History and Philosophy of Science* 32, no. 1 (2001): 1–29; and John Tresch, *The Romantic Machine: Utopian Science and Technology after Napoleon* (Chicago: University of Chicago Press, 2012), 270–72.

86. Toby A. Appel, "Henri de Blainville and the Animal Series: A Nineteenth-Century Chain of Being," *Journal of the History of Biology* 13, no. 2 (1980): 291–319, on 293.

87. See, e.g., Henri Ducrotay de Blainville, "Mémoire sur l'organisation d'une espèce de Mollusque nu de la famille des Limacinés," *Journal de Physique* 96 (1823): 175–87.

88. Spary, *Utopia's Garden*, 95.

89. See Jean-Louis-Maurice Laurent, "Expériences sur l'imbibition des tissus dans l'age embryonnaire," *Annales Françaises et Étrangères d'Anatomie et de Physiologie* 1 (1837), 81–84. On the Blainvillian orientation of the journal, of which Laurent was one of several coeditors, see Jean-Louis-Maurice Laurent, "Recherches sur les affinités et les differences naturelles des appareils des animaux," *Annales Françaises et Étrangères d'Anatomie et de Physiologie* 1 (1837): 221–39.

90. See, e.g., Alexander von Humboldt, *Essai sur la géographie des plantes: Accompagné d'un tableau physique des régions équinoxiales, fondé sur des mesures exécutées, depuis le dixième degré de latitude boréale jusqu'au dixième degré de latitude australe, pendant les années 1799, 1800, 1801, 1802 et 1803* (Paris: Levrault, Schoell et Compagnie, Libraires, 1805). See also Michael Dettelbach, "Humboldtian Science," in *Cultures of Natural History*, ed. Jardine, Secord, and Spary, 287–305.

91. Janet Browne, *The Secular Ark: Studies in the History of Biogeography* (New Haven, CT: Yale University Press, 1983), 52–53. See also Nils Güttler, *Das Kosmoskop: Karten und ihre Benutzer in der Pflanzengeographie des 19. Jahrhunderts* (Göttingen: Wallstein, 2014).

92. A. P. de Candolle, "Géographie botanique," in *Dictionnaire des sciences naturelles*, vol. 18, ed. F. G. Cuvier (Paris: Levrault, 1820), 359–436, on 384. Quoted (in slightly different translation) in Trevor Pearce, "'A Great Complication of Circumstances'—Darwin and the Economy of Nature," *Journal of the History of Biology* 43, no. 3 (2010): 493–528, on 503.

93. Candolle, "Géographie botanique," 421.

94. Isidore Geoffroy Saint-Hilaire, *Acclimatation et domestication des animaux utiles* (Paris: Librarie Agricole de la Maison Rustique, 1861), 501.

95. Geoffroy Saint-Hilaire, *Acclimatation et domestication des animaux utiles*, 242. See also Michael A. Osborne, *Nature, the Exotic, and the Science of French Colonialism* (Bloomington: Indiana University Press, 1994), 66–72.

96. See Charles Darwin, *On the Origin of Species by Means of Natural Selection, or the Preservation of Favoured Races in the Struggle for Life* (London: John Murray, 1859).

97. Geoffroy Saint-Hilaire, *Acclimatation et domestication des animaux utiles*, 517–19. See

also Rémi Luglia, *Des savants pour protéger la nature: La Société d'Acclimatation (1854–1960)* (Rennes: Presses Universitaires de Rennes, 2015).

98. Henri Ducrotay de Blainville and Francois-Louis-Michel Maupied, *Histoire des sciences de l'organisation et de leurs progrès, comme base de la philosophie*, vol. 1 (Paris: Jacques LeCoffre, 1847). See also Appel, "Henri de Blainville and the Animal Series"; John Tresch, "The Animal Series and the Genesis of Socialism," in *Of Elephants & Roses*, ed. Prince, 196–204; and Tresch, *Romantic Machine*, 269–75.

99. Henri Ducrotay de Blainville, introduction to Blainville and Maupied, *Histoire des sciences de l'organisation*, i–xxiii, quote on ii.

100. Comte, "Quarantième Leçon," 292–93. See also Tresch, *Romantic Machine*, 271.

101. See Harriet Martineau, *The Positive Philosophy of Auguste Comte* (London: John Chapman, 1853), 356–69; and Herbert Spencer, *The Principles of Psychology* (Longman, Brown, Green and Longmans, 1855), esp. chap. 5 ("The Correspondence Between Life and Its Circumstances"), 366–75. See also Trevor Pearce, "From 'Circumstances' to 'Environment': Herbert Spencer and the Origins of the Idea of Organism–Environment Interaction," *Studies in History and Philosophy of Biological and Biomedical Sciences* 41 (2010): 241–52, esp. 248.

102. Georges Cuvier, *Rapport historique sur les progrès des sciences naturelles depuis 1789, et sur leur état actuel* (Paris: Imprimerie Impériale, 1810), 152. See also Outram, "Uncertain Legislator," 333.

CHAPTER 2

1. *Army Medical Department Report for the Year 1867,* vol. 9 (London: Harrison and Sons, 1869), 76; H. Johnson, R. J. O'Flaherty, and Lieutenant Grain, "Report of the Commission Appointed to Assemble to Investigate the Origin, Progress, and Results of the Epidemic of Yellow Fever in the Island of Jamaica in 1866 and 1867," *Army Medical Department Report for the Year 1867*, appendix 3, 224–49.

2. *Army Medical Department Report for the Year 1867*, 9:76–77.

3. Johnson, O'Flaherty, and Grain, "Report of the Commission," 76.

4. Johnson, O'Flaherty, and Grain, 234.

5. Robert Lawson, "Observations on the Outbreak of Yellow Fever among the Troops at Newcastle, Jamaica, in the latter part of 1856," *British and Foreign Medico-Chirurgical Review: Or, Quarterly Journal of Practical Medicine and Surgery* 24 (1859): 324–49, on 325–26.

6. "Memoranda for the Guidance of the Commission of Enquiry into the Origins and Progress of the Recent Epidemic of Yellow Fever in Jamaica, Especially with Reference to Its Prevalence at Newcastle, 1867," in *Army Medical Department Report for the Year 1867,* 9:224–26, on 224.

7. "Memoranda for the Guidance of the Commission," 224–26.

8. Johnson, O'Flaherty, and Grain, "Report of the Commission," 226.

9. Hippocrates, *Ancient Medicine: Airs, Waters, Places*, trans. W. H. S. Jones, Loeb Classical Library 147 (Cambridge, MA: Harvard University Press, 1923), https://doi.org/10.4159/DLCL .hippocrates_cos-airs_waters_places.1923.

10. Mark Harrison, *Medicine in an Age of Commerce and Empire: Britain and Its Tropical Colonies, 1660–1830* (Oxford: Oxford University Press, 2010).

11. Sheldon Watts, "Yellow Fever Immunities in West Africa and the Americas in the Age of Slavery and Beyond: A Reappraisal," *Journal of Social History* 34, no. 4 (2001): 955–67. Now-discredited claims of an important role for inherent racial differences in immunity to tropical disease were common in earlier biological and environmental histories of European colonialism; see, e.g., Alfred W. Crosby, *The Columbian Exchange: Biological and Cultural Consequences of 1492* (Westport, CT: Greenwood, 1972); William H. McNeill, *Plagues and Peoples* (Garden City, NY: Anchor Press, 1977).

12. David S. Jones, "Virgin Soil Revisited," *The William and Mary Quarterly* 60, no. 4 (2003), 703–42.

13. J. R. McNeill, *Mosquito Empires: Ecology and War in the Greater Caribbean* (Cambridge: Cambridge University Press, 2010), 40; François Delaporte, *History of Yellow Fever: An Essay on the Birth of Tropical Medicine*, trans. Arthur Goldhammer (Cambridge, MA: MIT Press, 1991), 54–63.

14. Irving Rouse, *The Tainos: Rise and Decline of the People Who Greeted Columbus* (New Haven, CT: Yale University Press, 1992), 150–61.

15. Eric E. Williams, *Capitalism and Slavery* (Chapel Hill: University of North Carolina Press, 1944); Sidney W. Mintz, *Sweetness and Power: The Place of Sugar in Modern History* (New York: Viking, 1985).

16. J. R. McNeill, *Mosquito Empires*, 81–86.

17. On absentee white landowners in Jamaica, see Colleen A. Vasconcellos, *Slavery, Childhood, and Abolition in Jamaica, 1788–1838* (Georgia: University of Georgia Press, 2015), 2.

18. Christer Petley, *Slaveholders in Jamaica: Colonial Society and Culture during the Era of Abolition* (London: Pickering & Chatto, 2009), 6.

19. David Geggus, "Yellow Fever in the 1790s: The British Army in Occupied Saint Domingue," *Medical History* 29 (1979): 38–58; Michael Duffy, *Soldiers, Sugar, and Seapower: The British Expeditions to the West Indies and the War against Revolutionary France* (Oxford: Clarendon, 1987).

20. J. R. McNeill, *Mosquito Empires*, 247.

21. J. R. McNeill, 258–59.

22. J. R. McNeill, 254.

23. J. R. McNeill, 265.

24. Johnson, O'Flaherty, and Grain, "Report of the Commission," 248.

25. Philip D. Curtin, *Death by Migration: Europe's Encounter with the Tropical World in the Nineteenth Century* (New York: Cambridge University Press, 1989), 40–61.

26. Curtin, *Death by Migration*, 81–82.

27. Curtin, 18, 35.

28. Curtin, 4.

29. Peter Burroughs, "Imperial Institutions and the Government of Empire," in *Oxford History of the British Empire*, vol. 3 of The Nineteenth Century, ed. Andrew Porter (Oxford: Oxford University Press, 1999), 170–97.

30. Burroughs, "Imperial Institutions and the Government of Empire."

31. Mark Harrison, "'The Tender Frame of Man': Disease, Climate, and Racial Difference in India and the West Indies, 1760–1860," *Bulletin of the History of Medicine* 70, no. 1 (1996): 68–93, on 79; David Arnold, ed., *Warm Climates and Western Medicine: The Emergence of Tropical Medicine, 1500–1900* (Atlanta: Rodopi, 1996).

32. Curtin, *Death by Migration*, 3–4.

33. Thomas C. Holt, *The Problem of Freedom: Race, Labor, and Politics in Jamaica and Britain* (Baltimore: Johns Hopkins University Press, 1992); Mimi Sheller, *Democracy after Slavery: Black Publics and Peasant Radicalism in Haiti and Jamaica* (Miami: University Press of Florida, 2001).

34. Bernard Semmel, *The Governor Eyre Controversy* (London: MacGibbon & Kee, 1962); Gad J. Heuman, *The Killing Time: The Morant Bay Rebellion in Jamaica* (Knoxville: University of Tennessee Press, 1994).

35. Heuman, *The Killing Time*, 44–60.

36. See, e.g., Robert Jackson, *An Outline of the History and Cure of Fever, Endemic and Contagious* (Edinburgh: John Meir, 1808), 262–79.

37. Jackson, *Outline of the History and Cure of Fever*, 245–46.

38. A. M. Tulloch, "On the Sickness and Mortality among the Troops in the West Indies," part 3, *Journal of the Statistical Society of London* 1, no. 7 (1838): 428–44, on 433–34.

39. A. M. Tulloch, "On the Sickness and Mortality among the Troops in the West Indies," part 2, *Journal of the Statistical Society of London* 1, no. 4 (1838): 216–30, on 217.

40. Tulloch, *Statistical Report . . . in the West Indies*, 11.

41. Tulloch, "On the Sickness and Mortality among the Troops," part 2, 217.

42. Jackson, *Outline of the History and Cure of Fever*, 371.

43. John Hunter, *Observations on the Diseases of the Army in Jamaica: And on the Best Means of Preserving the Health of Europeans, in that Climate* (London: Printed for G. Nicol, Pall-Mall, Bookseller for His Majesty, 1788), 21–26. See also Mark Harrison, "Differences of Degree: Representations of India in British Medical Topography, 1820–c.1870," in *Medical Geography in Historical Perspective*, ed. Nicolaas A. Rupke (London: Wellcome Trust Centre for the History of Medicine at UCL, 2000), 51–69.

44. James M'Cabe, *Military-Medical Reports: Containing Pathological and Practical Observations Illustrating the Diseases of Warm Climates* (Cheltenham: G. A. Williams, 1825), 95.

45. M'Cabe, *Military-Medical Reports*, 94–99.

46. M'Cabe, 100–111.

47. On analogous hill stations in British India, see Dane Kennedy, *The Magic Mountains: Hill Stations and the British Raj* (Berkeley: University of California Press, 1996).

48. See *Report of Commissioners Appointed to Inquire into the Sanitary State of the Army in India, with Abstract of Evidence and of Reports Received from Indian Military Stations* (London: H. M. Stationery Office, 1864), 108.

49. William M. Gomm to Lord Fitzroy Somerset, December 26, 1840, in *The Story of Newcastle, Jamaica* (London: Vacher & Sons, 1864), 12–13, quote on 13.

50. Robert Armstrong, *The Influence of Climate and Other Agents on the Human Constitution, with Reference to the Causes and Prevention of Disease among Seamen: With Observations on Fever in General, and an Account of the Epidemic Fever of Jamaica* (London: Longman and Brown, 1843), x.

51. Curtin, *Death by Migration*, 3.

52. See Rupke, *Medical Geography in Historical Perspective*.

53. A. M. Tulloch, "On the Sickness and Mortality among the Troops," part 2, 219.

54. Ian Hacking, *The Taming of Chance* (Cambridge: Cambridge University Press, 1990); Gerd Gigerenzer, Zeno Swijtink, Theodore Porter, Lorraine Daston, John Beatty, and Lorenz Kruger, *The Empire of Chance: How Probability Changed Science and Everyday Life* (Cambridge: Cambridge University Press, 1989).

55. Theodore M. Porter, *The Rise of Statistical Thinking: 1820–1900* (Princeton, NJ: Princeton University Press, 1986), 23–25.

56. William Whewell, "On the Connexion of the Physical Sciences by Mrs. Somerville," *Quarterly Review* 51 (1834): 54–68. See also Richard R. Yeo, *Defining Science: William Whewell, Natural Knowledge and Public Debate in Early Victorian Britain* (New York: Cambridge University Press, 1993), 110–11.

57. Michel Foucault, *The History of Sexuality*, vol. 1, trans. Robert Hurley (New York: Pantheon, 1978), 133–60; Ian Hacking, "Biopower and the Avalanche of Printed Numbers," *Humanities in Society* 5 (1982): 279–95.

58. See, e.g., William Whewell, *On the Results of an Extensive System of Tide Observations Made on the Coasts of Europe and America in June 1835* (London: Richard Taylor, 1836). See also Michael S. Reidy, *Tides of History: Ocean Science and Her Majesty's Navy* (Chicago: University of Chicago Press, 2008), 157–97.

59. Catherine Kelly, *War and the Militarization of British Army Medicine, 1793–1830* (London: Routledge, 2011), 103–26.

60. James McGrigor, "Supplementary Chapter," in *The Autobiography and Services of Sir James McGrigor, Bart., Late Director-General of the Army Medical Department, with an Appendix of Notes and Original Correspondence* (London: Longman, Green, Longman and Roberts, 1861), 374–97.

61. Alexander M. Tulloch, introduction to *Statistical Report . . . in the West Indies*, iii–vi, on iii–iv.

62. Henry Marshall, *Notes on the Medical Topography of the Interior of Ceylon: And on the Health of the Troops Employed in the Kandyan Provinces, during the Years 1815, 1816, 1817, 1818, 1819, and 1820; with Brief Remarks on the Prevailing Diseases* (London: Burgess and Hill, 1821). See also Richard L. Blanco, "Henry Marshall (1775–1851) and the Health of the British Army," *Medical History* 14, no. 3 (1970): 260–76.

63. Tulloch, introduction to *Statistical Report . . . in the West Indies*, v.

64. See Tulloch, *Statistical Report . . . in the West Indies* (London: W. Clowes and Sons, 1838); Alexander M. Tulloch, *Statistical Reports on the Sickness, Mortality, and Invaliding among the Troops in the United Kingdom, the Mediterranean, and British America* (London: W. Clowes and Sons, 1839); and Alexander M. Tulloch, *Statistical Reports on the Sickness, Mor-*

tality, & Invaliding, among the Troops in Western Africa, St. Helena, the Cape of Good Hope, and the Mauritius (London: W. Clowes and Sons, 1840).

65. A. M. Tulloch, "On the Sickness and Mortality among the Troops in the West Indies," part 1, *Journal of the Statistical Society of London* 1, no. 3 (1838): 129–42, quote on 132.

66. See Andrew Halliday, *A Letter to the Right Honourable the Secretary at War, on Sickness and Mortality in the West Indies: Being a Review of Captain Tulloch's Statistical Report* (London: John W. Parker and Son, 1839).

67. Halliday, *Letter on Sickness and Mortality in the West Indies*, 3.

68. David Arnold, *Colonizing the Body: State Medicine and Epidemic Disease in Nineteenth-Century India* (Berkeley: University of California Press, 1993), 67.

69. Florence Nightingale, *Army Sanitary Administration and Reform under the Late Lord Herbert* (London: McCorquodale, 1862), 4.

70. Nightingale, *Army Sanitary Administration and Reform*, 6–7, quote on 6.

71. Edmund A. Parkes, *A Manual of Practical Hygiene: Prepared Especially for Use in the Medical Service of the Army* (London: John Churchill & Sons, 1864), 475.

72. Lawson, "Observations on the Outbreak of Yellow Fever among the Troops," 343.

73. See "Memoranda for the Guidance of the Commission," 224; and Johnson, O'Flaherty, and Grain, "Report of the Commission," 226.

74. See, e.g., Hunter, *Observations on the Diseases of the Army in Jamaica*, 22–23.

75. Linda Nash, *Inescapable Ecologies: A History of Environment, Disease, and Knowledge* (Berkeley: University of California Press, 2006), 27–29.

76. William Fergusson, "On the Nature and History of Marsh Poison," *Philadelphia Journal of the Medical and Physical Sciences* 7, no. 13 (1823): 585–97. (This is a reprint; Fergusson's essay was originally published as a pamphlet in 1821.)

77. Hunter, *Observations on Diseases of the Army in Jamaica*, 15. See also Harrison, *Medicine in an Age of Commerce and Empire*, 83–84.

78. Hunter, *Observations on Diseases of the Army in Jamaica*, 23.

79. A. M. Tulloch, "Statistical Report on the Sickness, Mortality, and Invaliding among the Troops Serving in British America," in *Statistical Reports . . . in the United Kingdom, the Mediterranean, and British America*, 1b–58b (separately paginated, with appendix), on 45b.

80. A. M. Tulloch, introduction to *Statistical Reports . . . in the United Kingdom, the Mediterranean, and British America* (1839), 3–4, on 4; T. Graham Balfour, "The Inaugural Address of Dr. T. Graham Balfour, F.R.S., &c., Honorary Physician to Her Majesty the Queen, President of the Royal Statistical Society. Delivered 20th November, 1888," *Journal of the Royal Statistical Society* 51, no. 4 (1888): 683–700, on 693. See also Curtin, *Death by Migration*, 43–45.

81. Harrison, *Medicine in an Age of Commerce and Empire*, 98.

82. Armstrong, *Influence of Climate and Other Agents*, 71.

83. Armstrong, ix.

84. Tulloch, "On Sickness and Mortality among the Troops," part 3, 429–30, 434–35.

85. Tulloch, 434.

86. Tulloch, 428.

87. A. M. Tulloch, "On the Mortality among Her Majesty's Troops Serving in the Colonies

during the Years 1844 and 1845," *Journal of the Statistical Society of London* 10, no. 3 (1847): 252-59, quote on 259.

88. Evelleen Richards, "The 'Moral Anatomy' of Robert Knox: The Interplay between Biological and Social Thought in Victorian Scientific Naturalism," *Journal of the History of Biology* 22 (1989): 373-436, esp. 379-80. See also Nancy Stepan, *The Idea of Race in Science: Great Britain, 1800-1960* (Hamden, CT: Archon Books, 1982); and George W. Stocking Jr., *Race, Culture, and Evolution: Essays in the History of Anthropology* (Chicago: University of Chicago Press, 1982).

89. Robert Knox, *The Races of Men: A Fragment* (Philadelphia: Lea & Blanchard, 1850), 77. For a similar citation of Tulloch's statistics by a prominent US race scientist, see Josiah C. Nott, "Acclimation: Or, the Comparative Influence of Climate, Endemic and Epidemic Diseases, on the Races of Men," in *Indigenous Races of the Earth*, ed. Josiah C. Nott and George R. Gliddon (Philadelphia: J. P. Lippincott, 1857), 353-401, on 374-75 and 382-86.

90. Trevor Pearce, "From 'Circumstances' to 'Environment': Herbert Spencer and the Origins of the Idea of Organism-Environment Interaction," *Studies in History and Philosophy of Biological and Biomedical Sciences* 41, no. 3 (2010): 241-52.

91. Knox, *Races of Men*, 76.

92. James Hunt, "On Ethno-climatology; or the Acclimatization of Man," *Transactions of the Ethnological Society of London* 2 (1863), 50-83, quote on 51. See also David N. Livingstone, "The Moral Discourse of Climate: Historical Considerations on Race, Place and Virtue," *Journal of Historical Geography* 17, no. 4 (1991): 413-34; and David N. Livingstone, "Race, Space and Moral Climatology: Notes toward a Genealogy," *Journal of Historical Geography* 28, no. 2 (2002): 159-80.

93. Hunt, "On Ethno-climatology," 52.

94. Johnson, O'Flaherty, and Grain, "Report of the Commission," 245. See also Curtin, *Death by Migration*, 70.

95. Johnson, O'Flaherty, and Grain, "Report of the Commission," 246.

96. Nancy Tomes, *The Gospel of Germs: Men, Women, and the Microbe in American Life* (Cambridge, MA: Harvard University Press, 1998); Michael Worboys, *Spreading Germs: Disease Theories and Medical Practice in Britain, 1865-1900* (New York: Cambridge University Press, 2000).

97. Robert Koch, "Die Ätiologie der Milzbrandkrankheit, begründet auf die Entwicklungsgeschichte des Bacillus Anthracis," in *Gesammelte Werke von Robert Koch*, vol. 1, ed. J. Schwalbe (1876; Leipzig: George Thieme, 1912), 5-26.

98. John Tyndall and Louis Pasteur, *Les microbes organisés, leur rôle dans la fermentation, la putréfaction et la contagion* (Paris: Gauthier-Villars, 1878); Claude Bernard, *Introduction à l'étude de la médecine expérimentale* (Paris: J. B. Baillière et Fils, 1865), 110. See also Frederic L. Holmes, "Claude Bernard, the 'Milieu Intérieur,' and Regulatory Physiology," *History and Philosophy of the Life Sciences* 8, no. 1 (1986): 3-25.

99. Patrick Manson, "On the Development of *Filaria sanguinis hominis*, and on the Mosquito Considered as a Nurse," *Journal of the Linnean Society of London, Zoology* 14, no. 75 (1878): 304-11.

100. Charles Finlay, "The Mosquito Hypothetically Considered as an Agent in the Transmission of Yellow Fever Poison," *New Orleans Medical and Surgical Journal* 9 (1881/82): 601–16, reprinted in *Yale Journal of Biological Medicine* 9, no. 6 (1937): 589–604. (Carlos Finlay's first name is sometimes written as Charles in English-language publications.) This paper and others are collected in *Trabajos Selectos del Dr. Carlos J. Finlay* (Habana: Secretaria de Sanidad y Beneficencia, 1912). See also Mariola Espinosa, *Epidemic Invasions: Yellow Fever and the Limits of Cuban Independence, 1878–1930* (Chicago: University of Chicago Press, 2009).

101. Finlay, "Mosquito Hypothetically Considered as an Agent," 590.

102. Charles Finlay, "Yellow Fever: Its Transmission by Means of the Culex Mosquito," *American Journal of the Medical Sciences* 184 (1886): 395–408; Finlay, "Mosquito Hypothetically Considered as an Agent," 601–2.

103. Finlay, "Yellow Fever," 401.

104. George Sternberg, "On Yellow Fever," in *Hygiene and Diseases of Warm Climates*, ed. Andrew Davidson (London: Young J. Pentland, 1893), 317. See also Espinosa, *Epidemic Invasions*, 57.

105. Sternberg, "On Yellow Fever," 288.

106. Walter Reed, "Recent Researches Concerning the Etiology, Propagation, and Prevention of Yellow Fever, by the United States Army Commission," *Journal of Hygiene* 2, no. 2 (1902): 101–19.

107. George M. Sternberg, "Sanitary Problems Connected with the Construction of the Isthmian Canal," *North American Review* 175, no. 550 (1902): 378–87, on 383.

108. Sternberg, "Sanitary Problems of the Isthmian Canal," 382.

109. Curtin, *Death by Migration*, 132; Margaret Jones, *Public Health in Jamaica, 1850–1940: Neglect, Philanthropy, and Development* (Kingston: University of the West Indies Press, 2013), 95–102.

110. Paul S. Sutter, "Nature's Agents or Agents of Empire? Entomological Workers and Environmental Change during the Construction of the Panama Canal," *Isis* 98, no. 4 (2007): 724–54, on 749.

111. "James Lane Notter, M.D.," *British Medical Journal* 2, no. 3279 (1923): 843–44.

112. J. Lane Notter, "Hygiene of the Tropics," in *Hygiene & Diseases of Warm Climates*, ed. Davidson, 25–82, on 74. See also Warwick Anderson, *Colonial Pathologies: American Tropical Medicine, Race, and Hygiene in the Philippines* (Durham, NC: Duke University Press, 2006), 24–25.

113. Notter, "Hygiene of the Tropics," 51.

114. Notter, 25. See also J. Lane Notter, "On the Sanitary Methods of Dealing with Epidemics," *Journal of the Military Service Institution of the United States* 25 (1899): 111–24.

115. Finlay, "Mosquito Hypothetically Considered as an Agent," 599.

116. Warwick Anderson, *Cultivation of Whiteness: Science, Health, and Racial Destiny in Australia* (Durham, NC: Duke University Press, 2006), 97; Paul S. Sutter, "The Tropics: A Brief History of an Environmental Imaginary," in *The Oxford Handbook of Environmental History*, ed. Andrew Isenberg (New York: Oxford University Press, 2017), 178–204, on 192.

117. Cf. E. M. Collingham, *Imperial Bodies: The Physical Experience of the Raj, c. 1800–1947* (Malden, MA: Blackwell, 2001), 3–4; Linda Nash, *Inescapable Ecologies: A History of Environment, Disease, and Knowledge* (Berkeley: University of California Press, 2006); and David S. Barnes, "'Until Cleansed and Purified': Landscapes of Health in the Interpermeable World," *Change over Time* 6, no. 2 (2016): 138–52.

118. Delaporte, *History of Yellow Fever*, 146.

119. Warwick Anderson, "Natural Histories of Infectious Disease: Ecological Vision in Twentieth-Century Biomedical Science," *Osiris* 10 (2004): 39–61. Contrast Jean-Baptiste Fressoz, "Circonvenir les circumfusa: La chimie, l'hygiénisme et la libéralisation des 'choses environnantes': France, 1750–1850," *Revue d'histoire moderne et contemporaine* 56, no. 4 (2009): 39–76.

120. Linda Nash, "Purity and Danger: Historical Reflections on the Regulation of Environmental Pollutants," *Environmental History* 13, no. 4 (2008): 651–58, quote on 651.

121. "Memoranda for the Guidance of the Commission," 224.

CHAPTER 3

1. Jane Addams, *Twenty Years at Hull-House: With Autobiographical Notes* (New York: Macmillan, 1910), 92–93.

2. Addams, *Twenty Years at Hull-House*, 93.

3. Jane Addams, "The Objective Value of a Social Settlement," in Jane Addams, Robert A. Woods, J. O. S. Huntington, Franklin H. Giddings, and Bernard Bosanquet, *Philanthropy and Social Progress: Seven Essays* (New York: Thomas Y. Crowell, 1893), 27–56, quote on 29.

4. James Linn, *Jane Addams: A Biography* (New York: Appelton-Century, 1935), 65–90; Louise W. Knight, *Citizen: Jane Addams and the Struggle for Democracy* (Chicago: University of Chicago Press, 2005), 139–55.

5. Jane Addams, "The Subjective Necessity for Social Settlements," in Addams et al., *Philanthropy and Social Progress*, 1–26, on 2, 4.

6. Views of the exposition can be found in *The Columbian Exposition Album* (Chicago: Rand, McNally, 1893). On the symbolism of the exposition's spectacular displays of electric light, see David Nye, *Electrifying America: Social Meanings of a New Technology, 1880–1940* (Cambridge, MA: MIT Press, 1990), 37–43.

7. Carlos C. Closson Jr., "The Unemployed in American Cities," *Quarterly Journal of Economics* 8, no. 2 (1894): 168–217, on 189.

8. Addams, *Twenty Years at Hull-House*, 159–60.

9. William Cronon, *Nature's Metropolis: Chicago and the Great West* (New York: W. W. Norton, 1991).

10. Cronon, *Nature's Metropolis*, 56.

11. "The Position of the Horse in Modern Society," *Nation* 15 (October 31, 1872): 278. See also Ann Norton Greene, *Horses at Work: Harnessing Power in Industrial America* (Cambridge, MA: Harvard University Press, 2008), 1–9.

12. W. Hamilton Gibson, "Foreground and Vista at the Fair," *Scribner's Magazine*, July 1893,

29–37, quote on 31. See also Victoria Post Ranney, *Olmsted in Chicago* (Chicago: R. R. Donnelly, 1972).

13. Colin Fisher, *Urban Green: Nature, Recreation, and the Working Class in Industrial Chicago* (Chapel Hill: University of North Carolina Press, 2015), 9–16.

14. Addams, *Twenty Years at Hull-House*, 177–78. See also Knight, *Citizen: Addams and Democracy*, 151.

15. Theda Skocpol, *Protecting Soldiers and Mothers: The Political Origins of Social Policy in the United States* (Cambridge, MA: Harvard University Press, 1992).

16. See, e.g., Char Miller, *Gifford Pinchot and the Making of Modern Environmentalism* (Washington, DC: Island Press/Shearwater Books, 2001); and Donald Worster, *A Passion for Nature: The Life of John Muir* (New York: Oxford University Press, 2008).

17. Charles Booth, *Labor and Life of the People*, vol. 1 (London: Williams and Norgate, 1889); Addams, *Twenty Years at Hull-House*, 68.

18. Kevin Bales, "Charles Booth's *Survey of Life and Labour of the People in London 1889–1903*," in *The Social Survey in Historical Perspective, 1880–1940*, ed. Martin Bulmer, Kevin Bales, and Kathryn Kish Sklar (New York: Cambridge University Press, 1991), 66–110.

19. Auguste Comte, *Cours de Philosophie Positive*, vol. 4 (Paris: Bachelier, 1839), 252.

20. For the use of the terms *enumerator* and *schedule*, see Robert Hunter, *Tenement Conditions in Chicago: Report by the Investigating Committee of the City Homes Association* (Chicago: City Homes Association, 1901), 67, 202.

21. Kathryn Kish Sklar, *Florence Kelley and the Nation's Work* (New Haven, CT: Yale University Press, 1995), 230.

22. Residents of Hull-House, *Hull-House Maps and Papers* (New York: Thomas Y. Crowell, 1895). See also Sklar, *Florence Kelley*, 206–36.

23. See W. E. B. Du Bois and Isabel Eaton, *The Philadelphia Negro: A Social Study* (Philadelphia: University of Pennsylvania, 1899). See also Michael B. Katz and Thomas J. Sugrue, eds., *W. E. B. Du Bois, Race, and the City: "The Philadelphia Negro" and Its Legacy* (Philadelphia: University of Pennsylvania Press, 1998); and Aldon D. Morris, *The Scholar Denied: W. E. B. Du Bois and the Birth of Modern Sociology* (Oakland: University of California Press, 2015).

24. Paul Underwood Kellogg, ed., *The Pittsburgh Survey: Findings in Six Volumes*, vol. 1: Elizabeth Beardsley Butler, *Women and the Trades* (New York: New York Charities Publication Committee, 1909); vol. 2: Crystal Eastman, *Work-Accidents and the Law* (New York: New York Charities Publication Committee, 1910); vol. 3: John A. Fitch, *The Steel Workers* (New York: New York Charities Publication Committee, 1911); vol. 4: Margaret F. Byington, *Homestead: The Households of a Mill Town* (New York: New York Charities Publication Committee, 1911); vol. 5: Paul Underwood Kellogg, ed., *The Pittsburgh District: Civic Frontage* (New York: New York Survey Associates, 1914); and vol. 6: Paul Underwood Kellogg, ed., *Wage-Earning Pittsburgh* (New York: New Survey Associates, 1914).

25. Joel Tarr, *The Search for the Ultimate Sink: Urban Pollution in Historical Perspective* (Akron, OH: University of Akron Press, 1996), 77–102.

26. Hunter, *Tenement Conditions in Chicago*.

27. Hunter, *Tenement Conditions in Chicago*, 36–41.

28. Hunter, 49.

29. See Florence Kelley and Alzina P. Stevens, "Wage-Earning Children," in *Hull-House Maps and Papers*, 49–79; and Friedrich Engels, *The Condition of the Working Class in England*, trans. Florence Kelley Wischnewetzky (London: George Allen & Unwin, 1892). See also Sklar, *Florence Kelley*, esp. 171–205.

30. See Alice Hamilton, "Lead Poisoning in Illinois," *American Economic Review* 1, no. 2 (1911): 257–64.

31. Alice Hamilton, *Exploring the Dangerous Trades* (Boston: Little, Brown, 1943). See also Robert Gottlieb, *Forcing the Spring: The Transformation of the American Environmental Movement*, rev. ed. (Washington, DC: Island Press, 2005), 83–88; Christopher C. Sellers, *Hazards of the Job: From Industrial Disease to Environmental Health Science* (Chapel Hill: University of North Carolina Press, 1997), 69–140; and Barbara Sicherman, "Working It Out: Gender, Profession, and Reform in the Career of Alice Hamilton," in *Gender, Class, Race, and Reform in the Progressive Era*, ed. Noralee Frankel and Nancy S. Dye (Lexington: University Press of Kentucky, 2015), 127–47.

32. Byington, *Homestead: Households of a Mill Town*, photographs on plates between 28–29 and 120–12; Hunter, *Tenement Conditions in Chicago*, 63, 139.

33. Donald Pizer, "The Problem of American Literary Naturalism and Theodore Dreiser's 'Sister Carrie,'" *American Literary Realism* 32, no. 1 (1999): 1–11.

34. See Theodore Dreiser, *Sister Carrie* (New York: Doubleday, Page, 1900); and Upton Sinclair, *The Jungle* (New York: Doubleday, Page, 1906).

35. Albion W. Small and George E. Vincent, *An Introduction to the Study of Sociology* (New York: American Book Company, 1894), 165–66, quote on 166. See also Dorothy Ross, *The Origins of American Social Science* (New York: Cambridge University Press, 1991), 226; and Mary Jo Deegan, *Jane Addams and the Men of the Chicago School, 1892–1918* (New Brunswick, NJ: Transaction Books, 1988), 54–69.

36. E. g., Dorothea Moore, "A Day at Hull House," *American Journal of Sociology* 2, no. 5 (1897): 629–42; Grace Abbott, "A Study of the Greeks in Chicago," *American Journal of Sociology* 15, no. 3 (1909): 379–93; Jane Addams, "A Belated Industry," *American Journal of Sociology* 1, no. 5 (1896): 536–50; Jane Addams, "Trades Unions and Public Duty," *American Journal of Sociology* 4, no. 4 (1899): 448–62; Jane Addams, "Problems of Municipal Administration," *American Journal of Sociology* 10, no. 4 (1905): 425–44.

37. Addams, *Twenty Years at Hull-House*, 94.

38. Addams, "Objective Value," 47; Addams, *Twenty Years at Hull-House*, 101–6.

39. "Hull-House: A Social Settlement," in *Hull-House Maps and Papers*, 207–30, esp. 223–24; Addams, *Twenty Years at Hull-House*, 438–39.

40. Addams, *Twenty Years at Hull-House*, 237.

41. Addams, 315–22.

42. Louise W. Knight, "Garbage and Democracy: The Chicago Community Organizing Campaign of the 1890s," *Journal of Community Practice* 14, no. 3 (2007): 7–27.

43. Sklar, *Florence Kelley*, 237; Hamilton, *Exploring the Dangerous Trades*, 135–36; Sellers, *Hazards of the Job*, 70–71.

44. Addams, *Twenty Years at Hull-House*, 315–17. See also Harold L. Platt, "Jane Addams

and the Ward Boss Revisited: Class, Politics, and Public Health in Chicago, 1890–1930," *Environmental History* 5, no. 2 (2000): 194–22.

45. "Tots in Gay Frolic," *Chicago Daily Tribune*, June 9, 1894; Addams, *Twenty Years at Hull-House*, 289–92. See also Paul S. Boyer, *Urban Masses and Moral Order in America, 1820–1920* (Cambridge, MA: Harvard University Press, 1978), 233–51; and Dominick Cavallo, *Muscles and Morals: Organized Playgrounds and Urban Reform, 1880–1920* (Philadelphia: University of Pennsylvania Press, 1981).

46. Daniel T. Rodgers, *Atlantic Crossings: Social Politics in a Progressive Age* (Cambridge, MA: Belknap Press of Harvard University Press, 1998), 53–54.

47. Helen Meller, *Patrick Geddes: Social Evolutionist and City Planner* (New York: Routledge, 1990), 52.

48. Meller, *Patrick Geddes*, 81. Addams was aware of Geddes's work before his visit; see, e.g., Jane Addams, "A Function of the Social Settlement," *Annals of the American Academy of Political and Social Science* 13 (1899): 33–55, on 48.

49. Daniel H. Burnham and Edward H. Bennett, *Plan of Chicago* (Chicago: Commercial Club, 1909), plate LXXXV, between 80 and 81. See also Boyer, *Urban Masses and Moral Order*, 270–76.

50. Addams, *Twenty Years at Hull-House*, 81.

51. Addams, 80. See also Beth Eddy, "The Cathedral of Humanity on Halsted Street: Jane Addams, Auguste Comte, and Edward Caird," *Soundings: An Interdisciplinary Journal* 97, no. 1 (2014): 1–20.

52. Addams, "A Function of the Social Settlement," 47; Jane Addams, *Democracy and Social Ethics* (New York: Macmillan, 1902), 173.

53. Herbert Spencer, *The Principles of Sociology*, 2nd ed., vol. 1 (London: Williams and Norgate, 1877), 39. On Spencer's influence in the United States, see Richard Hofstadter, *Social Darwinism in American Thought, 1860–1915* (Philadelphia: University of Pennsylvania Press, 1944); and William F. Fine, *Progressive Evolutionism and American Sociology, 1890–1920* (Ann Arbor, MI: UMI Research Press, 1979).

54. Lester F. Ward, "Contributions to Social Philosophy: VIII. The Mechanics of Society," *American Journal of Sociology* 2, no. 2 (1896): 234–54, definitions of *genetic* and *telic* on 246–48.

55. For "accelerate social evolution," see Lester F. Ward, "Contributions to Social Philosophy: IX. The Purpose of Sociology," *American Journal of Sociology* 2, no. 3 (1896): 446–60, quote on 457. For "conscious improvement," see the title of Lester F. Ward, *Applied Sociology: A Treatise on the Conscious Improvement of Society by Society* (Boston: Ginn, 1906).

56. Small and Vincent, *Introduction to the Study of Society*, 57.

57. Small and Vincent, *Introduction to the Study of Society*, 87.

58. Small and Vincent, 90–91.

59. Small and Vincent, 89. Small and Vincent drew this definition from J. S. Mackenzie, *An Introduction to Social Philosophy* (Glasgow: James MacLehose & Sons, 1890), 238.

60. Deegan, *Addams and the Men of the Chicago School*, 80–83, 249–53.

61. John Dewey, "Evolution and Ethics," *Monist* 8, no. 3 (1898): 321–41. See also Beth L. Eddy, *Evolutionary Pragmatism and Ethics* (Lanham, MD: Lexington Books, 2016); and Trevor

Pearce, "American Pragmatism, Evolution, and Ethics," in *The Cambridge Handbook of Evolutionary Ethics*, ed. Michael Ruse and Robert J. Richards (Cambridge: Cambridge University Press, 2017), 43–57. On Addams's adoption of pragmatism and her deployment of evolutionary theory, see Knight, *Citizen: Addams and Democracy*, 357–58.

62. Jane Addams, *Newer Ideals of Peace* (New York: Macmillan, 1906), 32–33.

63. Charlotte Perkins Gilman, *Women and Economics: A Study of the Economic Relation between Men and Women as a Factor in Social Evolution* (Boston: Small, Maynard, 1898). See also Mary A. Hill, *Charlotte Perkins Gilman: The Making of a Radical Feminist, 1860–1896* (Philadelphia: Temple University Press, 1980), 272–82. On Ward's influence on Gilman, see Sklar, *Florence Kelley*, 305.

64. Gilman, *Women and Economics*, 68.

65. Gilman, 318.

66. Gilman, vii. See also Kimberly A. Hamlin, *From Eve to Evolution: Darwin, Science, and Women's Rights in Gilded Age America* (Chicago: University of Chicago Press, 2014), 118–19.

67. W. E. Burghardt Du Bois, "The Study of the Negro Problems," *Annals of the American Academy of Political and Social Science* 11 (1898): 1–23, quote on 19.

68. Du Bois, *Philadelphia Negro*, 5.

69. Du Bois, 98.

70. Gilman, *Women and Economics*, vii.

71. Addams, *Democracy and Social Ethics*, 256.

72. Addams, "A Function of the Social Settlement," 36.

73. Addams, 36.

74. Addams, *Twenty Years at Hull-House*, 291.

75. Allen F. Davis, *Spearheads for Reform: The Social Settlements and the Progressive Movement, 1890–1914* (New York: Oxford University Press, 1967), 228–32.

76. Mina Carson, *Settlement Folk: Social Thought and the American Settlement Movement, 1885–1930* (Chicago: University of Chicago Press, 1990), 182–98.

77. The number declined from over 400 in 1910 to about 230 in the 1930s; see Stanley Wenocur and Michael Reisch, *From Charity to Enterprise: The Development of American Social Work in a Market Economy* (Urbana: University of Illinois Press, 1989), 140–41.

78. Peggy Glowacki and Julia Hendry, *Hull-House* (Chicago: Arcadia, 2004), 91–102.

79. Deegan, *Addams and the Men of the Chicago School*, 152–55. See also Zine Magubane, "Science, Reform, and the 'Science of Reform': Booker T Washington, Robert Park, and the Making of a 'Science of Society,'" *Current Sociology Monograph* 62, no. 4 (2014): 568–83; Fred H. Matthews, *Quest for an American Sociology: Robert E. Park and the Chicago School* (Montreal: McGill-Queen's University Press, 1977), 79; Pierre Lannoy, "When Robert E. Park Was (Re)Writing 'The City': Biography, the Social Survey, and the Science of Sociology," *American Sociologist* 35, no. 1 (2004): 34–62; and Boyer, *Urban Masses and Moral Order*, 286–87.

80. Vicky M. MacLean and Joyce E. Williams, "'Ghosts of Sociologies Past': Settlement Sociology in the Progressive Era at the Chicago School of Civics and Philanthropy," *American Sociologist* 43, no. 3 (2012): 235–63.

81. Quoted in Martin Bulmer, *The Chicago School of Sociology: Institutionalization, Diversity, and the Rise of Sociological Research* (Chicago: University of Chicago Press, 1984), 97.

82. "Committee's Preface," in *The Hobo: The Sociology of the Homeless Man*, by Nels Anderson (Chicago: University of Chicago Press, 1923), ix–xi, on x. See also Robert E. Kohler, *Inside Science: Stories from the Field in Human and Animal Science* (Chicago: University of Chicago Press, 2019), 59–92.

83. For "definition of the situation," see W. I. Thomas, "The Comparative Study of Cultures," *American Journal of Sociology* 42, no. 2 (1936): 177–85, on 184–85. See also William Isaac Thomas and Florian Znaniecki, "Methodological Note," in *The Polish Peasant in Europe and America: Monograph of an Immigrant Group*, by Thomas and Znaniecki, ed. Eli Zaretsky, vol. 1 (Chicago: University of Chicago Press, 1918), 1–86.

84. Addams, *Twenty Years at Hull-House*, 308. See also Thomas F. Gieryn, "City as Truth-Spot: Laboratories and Field Sites in Urban Studies," *Social Studies of Science* 36, no. 1 (2006): 5–38; and B. Robert Owens, "'Laboratory Talk' in U.S. Sociology, 1890–1930: The Performance of Scientific Legitimacy," *Journal of the History of the Behavioral Sciences* 50, no. 3 (2014): 302–20.

85. See Anderson, *The Hobo*; William Isaac Thomas, *The Unadjusted Girl: With Cases and Standpoint for Behavior Analysis* (Boston: Little, Brown, 1923); Frederic M. Thrasher, *The Gang: A Study of 1,313 Gangs in Chicago* (Chicago: University of Chicago Press, 1927); Harvey Warren Zorbaugh, *The Gold Coast and Slum: A Sociological Study of Chicago's Near North Side* (Chicago: University of Chicago Press, 1929); Clifford R. Shaw, *The Jack-Roller: A Delinquent Boy's Own Story* (Chicago: University of Chicago Press, 1930). See also Bulmer, *Chicago School of Sociology*, 45–63.

86. Ernest W. Burgess, "Discussion," in Shaw, *Jack-Roller*, 184–97, quote on 197.

87. Anderson's study was financed by the Committee on Homeless Men established by the Executive Committee of the Chicago Council of Social Agencies, with support from the Juvenile Protective Association; "Committee's Preface" in Anderson, *Hobo*, ix.

88. Robert E. Park, editor's preface to Anderson, *Hobo*, v–viii, quote on viii. See also Daniel Breslau, "The Scientific Appropriation of Social Research: Robert Park's Human Ecology and American Sociology," *Theory and Society* 19, no. 4 (1990): 417–46.

89. Ross, *Origins of American Social Science*, 363–64.

90. Robert E. Park and Ernest W. Burgess, *The City* (Chicago: University of Chicago Press, 1925), 145.

91. For "moral climate," see Robert E. Park, "The City: Suggestions for the Investigation of Human Behavior in the City Environment," *American Journal of Sociology* 20, no. 5 (1915): 577–612, on 608. For "social environment," see Robert E. Park, "Community Organization and Juvenile Delinquency," in Park et al., *The City*, 99–112, on 100–101.

92. Ross, *Origins of American Social Science*, 390–470.

93. Park, "Community Organization and Juvenile Delinquency," 111.

94. Park, 111.

95. See Robert E. Park, "The Natural History of the Newspaper," *American Journal of Sociology* 29, no. 3 (1923): 273–89.

96. Park, "The City: Suggestions for the Investigation of Human Behavior in the Urban Environment," in Park and Burgess, *The City*, 1–46, quote on 39.

97. Park, "The City," in Park and Burgess, *The City*, 22, 28.

98. Walter Lippmann, *Public Opinion* (New York: Harcourt, Brace, 1922), 15. For a detailed discussion of Lippmann's arguments as a response to John Dewey's pragmatic philosophy of democracy, see also Robert B. Westbrook, *John Dewey and American Democracy* (Ithaca, NY: Cornell University Press, 1991), 275–318.

99. Lippmann, *Public Opinion*, 29.

100. Lippmann, 29–32. For Dewey's response to Lippmann, see John Dewey, *The Public and Its Problems* (New York: H. Holt, 1927).

101. Lippmann, *Public Opinion*, 15–16.

102. Addams, "Subjective Necessity for Social Settlements," 22–23. The importance of Hull House's capacity to change was also stressed by resident Dorothea Moore, who described it as "a thing whose being is essentially plastic" (i.e., moldable); Moore, "A Day at Hull House," 630.

103. Addams, "Subjective Necessity for Social Settlements," 22–23.

CHAPTER 4

1. Ronald H. Limbaugh, *Tungsten in Peace and War, 1918–1946* (Reno: University of Nevada Press, 2010), 18–42.

2. "Wolfram Ore and Tungsten," *Journal of the Royal Society of Arts* 66, no. 3436 (1918): 702–3, on 702. See also Kendall E. Bailes, *Science and Russian Culture in an Age of Revolutions: V. I. Vernadsky and His Scientific School, 1863–1945* (Bloomington: Indiana University Press, 1990), 138–39.

3. Alfred G. White, "Economic Aspects of the World Mineral Situation," *Annals of the American Academy of Political and Social Science* 83 (1919): 70–85, on 76.

4. W. I. Vernadsky, "The Biosphere and the Noösphere," *American Scientist* 33, no. 1 (1945): 1–12, on 5; Bailes, *Science and Russian Culture*, 138–40, Alexei Kojevnikov, "The Great War, the Russian Civil War, and the Invention of Big Science," *Science in Context* 15, no. 2 (2002): 239–75, on 251–54; Jonathan D. Oldfield and Denis J. B. Shaw, "V. I. Vernadskii and the Development of Biogeochemical Understandings of the Biosphere, c.1880s–1968," *British Journal for the History of Science* 46, no. 2 (2013): 287–310; Alexei B. Kozhevnikov, *Stalin's Great Science: The Times and Adventures of Soviet Physicists* (London: Imperial College Press, 2004), 17–22.

5. Bailes, *Science and Russian Culture*, 138. Because the transliteration of Cyrillic names has varied across languages and changed over time, Vernadskii is often spelled Vernadsky, and Vladimir is sometimes spelled Wladimir.

6. Vladimir I. Vernadskii, "Ob izuchenii yestestvennykh proizvoditel'nykh sil Rossii" [On the study of the natural productive forces of Russia], *Ocherki i rechi* [Essays and speeches], vol. 1 (1922), 1–25, quote on 5. Quoted and translated in Kojevnikov, "Great War, Russian Civil War, and Big Science," 252.

7. Heinz Kautzleben and Axel Müller, "Vladimir Ivanovich Vernadsky (1863–1945)—from Mineral to Noosphere," *Journal of Geochemical Exploration* 147 (2014): 4–10, on 7.

8. Richard P. Tucker, *Insatiable Appetite: The United States and the Ecological Degradation*

of the Tropical World (Berkeley: University of California Press, 2000), 248. See also Richard P. Tucker, Tait Keller, J. R. McNeill, and Martin Schmid, eds., *Environmental Histories of the First World War* (Cambridge: Cambridge University Press, 2018).

9. For the United States, see Edson S. Bastin, "War-Time Mineral Activities in Washington," *Economic Geology* 13 (1918): 524–37; "The Joint Information Board on Minerals and Derivatives," *Science* 47, no. 1216 (1918): 385–86; and Whitman Cross, "Geology in the World War and After," *Bulletin of the Geological Society of America* 30 (1919): 165–88. See also Walter E. Pittman, "American Geologists at War: World War I," *Reviews in Engineering Geology*, vol. 13: *Military Geology in War and Peace*, ed. James R. Underwood and Peter L. Guth (Boulder, CO: Geological Society of America, 1998), 41–47.

10. Hew Strachan, "From Cabinet War to Total War: The Perspective of Military Doctrine, 1861–1918," in *Great War, Total War: Combat and Mobilization on the Western Front, 1914–1918*, ed. Roger Chickering and Stig Förster (Cambridge: Cambridge University Press, 2000), 19–34.

11. Gregory T. Cushman, *Guano and the Opening of the Pacific World: A Global Ecological History* (New York: Cambridge University Press, 2013), 156.

12. United States Food Administration, *Ten Lessons in Food Conservation* (Washington, DC: Government Printing Office, 1917), 3.

13. *An Act to Provide Further for the National Security and Defense by Encouraging the Production, Conserving the Supply, and Controlling the Distribution of Food Products and Fuel*, Public Law 41, *U.S. Statutes at Large* 40 (1917): 276–87; United States Food Commission, "Food for All—a Fundamental War Problem," *Scientific American* 118, no. 14 (1918): 310–11; William Clinton Mullendore, *History of the United States Food Administration, 1917–1919* (Stanford, CA: Stanford University Press, 1941). See also Timothy Johnson, "Nitrogen Nation: The Legacy of World War I and the Politics of Chemical Agriculture in the United States, 1916–1933," *Agricultural History* 90, no. 2 (2016): 209–29.

14. D. F. Houston, "Report of the Secretary of Agriculture," in *Annual Reports of the Department of Agriculture for the Year Ended June 30, 1919* (Washington, DC: Government Printing Office, 1920), 3–46, on 4; see also tables on 6–7.

15. "German Destruction in Northern France," *Iron Age* 104, no. 2 (July 10, 1919): 85–90.

16. George A. L. Dumont, "Devastation of the French Coal Mines," *Military Engineer* 15, no. 84 (1923): 487–94, on 489.

17. William H. Scheifley, "The Depleted Forests of France," *North American Review* 212, no. 778 (1920): 378–86, quote on 379.

18. *Final Report of the Forestry Sub-committee of the Reconstruction Committee* (London: H. M. Stationery Office, 1918), 23–34.

19. F. D. Acland, "The Prospects of Starting State Forestry," *Contemporary Review* 115 (1919): 386–95, on 387–88.

20. Edward Percy Stebbing, *British Forestry: Its Present Position and Outlook after the War* (London: John Murray, 1916), 29.

21. *Final Report of Forestry Sub-committee*, 46.

22. A. Joshua West, "Forests and National Security: British and American Forestry Policy in the Wake of World War I," *Environmental History* 8, no. 2 (2003): 270–93, on 275. Forests

in France were similarly devastated; see Chris Pearson, *Mobilizing Nature: The Environmental History of War and Militarization in Modern France* (Manchester: Manchester University Press, 2016), 92; and Jean-Yves Puyo, "Les Conséquences de la Première Guerre Mondiale pour les Forêts et les Forestiers Français," *Revue Forestière Française* 56 (2004): 573–84.

23. Vaclav Smil, *Enriching the Earth: Fritz Haber, Carl Bosch, and the Transformation of World Food Production* (Cambridge, MA: MIT Press, 2004).

24. Smil, *Enriching the Earth*, 103–8.

25. Thomas Parke Hughes, "Technological Momentum in History: Hydrogenation in Germany, 1898–1933," *Past & Present*, no. 44 (1969): 106–32, on 110.

26. William Notz, "The World's Coal Situation during the War: I," *Journal of Political Economy* 26, no. 6 (1918): 567–611, on 568.

27. Hugh S. Gorman, *The Story of N: A Social History of the Nitrogen Cycle and the Challenge of Sustainability* (New Brunswick, NJ: Rutgers University Press, 2013), 84–97.

28. Deborah K. Fitzgerald, *Every Farm a Factory: The Industrial Ideal in American Agriculture* (New Haven, CT: Yale University Press, 2003), 17–20.

29. *First Annual Report of the Forestry Commissioners, Year Ending September 30th, 1920* (London: H. M. Stationery Office, 1921), 11–15; E. P. Stebbing, "The Forestry Commission: The First Twenty-Five Years," *Nature* 155, no. 3933 (March 17, 1945): 317–18.

30. George Otis Smith, ed., *The Strategy of Minerals: A Study of the Mineral Factor in the World Position of America in War and in Peace* (New York: D. Appleton, 1919). See also Tucker et al., *Environmental Histories of the First World War*.

31. Andrea Westermann, "Geology and World Politics: Mineral Resource Appraisals as Tools of Geopolitical Calculation, 1919–1939," *Historical Social Research/Historische Sozialforschung* 40, no. 2 (2015): 151–73, on 159.

32. W. Stanley Jevons, *The Coal Question: An Enquiry Concerning the Progress of the Nation, and the Probable Exhaustion of Our Coal-Mines* (London: Macmillan, 1865), 305–27. See also Fredrik Albritton Jonsson, *Enlightenment's Frontier: The Scottish Highlands and the Origins of Environmentalism* (New Haven, CT: Yale University Press, 2013), 186.

33. Alfred G. White, "Economic Aspects of the World Mineral Situation," *Annals of the American Academy of Political and Social Science* 83 (1919): 70–85, on 78; Joseph E. Pogue, "Mineral Resources in War and Their Bearing on Preparedness," *Scientific Monthly* 5, no. 2 (1917): 120–34, on 127.

34. Bailes, *Science and Russian Culture*, 76–77.

35. Andy Bruno, "A Eurasian Mineralogy: Aleksandr Fersman's Conception of the Natural World," *Isis* 107, no. 3 (2016): 518–39, on 523–34.

36. Kendall E. Bailes, *Technology and Society under Lenin and Stalin: Origins of the Soviet Technical Intelligentsia, 1917–1941* (Princeton, NJ: Princeton University Press, 1978), 41–43.

37. V. M. Goldschmidt, "Teknisk-videnskabelig forskningsarbeide i utlandet og i Norge," *Samtiden* 29 (1918): 619–29, on 622.

38. Brian Mason, *Victor Moritz Goldschmidt: Father of Modern Geochemistry*, Special Publication Number 4 (San Antonio, TX: Geochemical Society, 1992), 23–24.

39. V. M. Goldschmidt, "Olivine and Forsterite Refractories in Europe," *Industrial and Engineering Chemistry* 30, no. 1 (1938): 32–34. See also Mason, *Victor Moritz Goldschmidt*, 25–26.

40. Mason, *Victor Moritz Goldschmidt*, 25. Despite Goldschmidt's efforts, extracting magnesium from olivine remained more expensive than extracting it from seawater.

41. West, "Forests and National Security," 278.

42. Henry E. Lowood, "The Calculating Forester: Quantification, Cameral Science, and the Emergence of Scientific Forestry Management in Germany," in *The Quantifying Spirit in the 18th Century*, ed. Tore Frängsmyr, J. L. Heilbron, and Robin E. Rider (Berkeley: University of California Press, 1990), 315-42.

43. Edward Percy Stebbing, "Forestry and the War," *Journal of the Royal Society of Arts* 64, no. 3304 (March 17, 1916): 350-60, quote on 359.

44. Miles Menander Dawson, "The Dynamics of Mobilization of Human Resources," *Annals of the American Academy of Political and Social Science* 78 (1918): 7-15.

45. Robert M. Yerkes, *Psychological Examining in the United States Army* (Washington, DC: US Government Printing Office, 1921); E. K. Strong, "Work of the Committee on Classification of Personnel in the Army," *Journal of Applied Psychology* 2, no. 2 (1918): 130-39. See also Daniel J. Kevles, "Testing the Army's Intelligence: Psychologists and the Military in World War I," *Journal of American History* 55 (1968): 565-81; and John Carson, *The Measure of Merit: Talents, Intelligence, and Inequality in the French and American Republics, 1750-1940* (Princeton, NJ: Princeton University Press, 2007), 197-219.

46. Lotus D. Coffman, "The Rehabilitation of Disabled Soldiers," *Journal of Education* 89, no. 12 (1919): 327-29, on 328. See also Beth Linker, *War's Waste: Rehabilitation in World War I America* (Chicago: University of Chicago Press, 2011), 153-54.

47. Vernadsky, "The Biosphere and the Noösphere," 5. See also Paul Josephson and Thomas Zeller, "The Transformation of Nature under Hitler and Stalin," in *Science and Ideology: A Comparative History*, ed. Mark Walker (New York: Routledge, 2003), 124-55.

48. Ernst Jünger, "Die totale Mobilmachung," in *Krieg und Krieger*, ed. Ernst Jünger (Berlin: Junker und Dünnhaupt Verlag, 1930), 9-30, reprinted in *Sämtliche Werke, Band 8: Essays I: Betrachtungen der Zeit* (Stuttgart: Klett-Cotta, 1980), 119-42. See also Roger Chickering, "Sore Loser: Ludendorff's Total War," in *The Shadows of Total War: Europe, East Asia, and the United States, 1919-1939*, ed. Roger Chickering and Stig Forster (Washington, DC: German Historical Institute and New York: Cambridge University Press, 2003), 151-66.

49. Mark Mazower, *Governing the World: The History of an Idea, 1815 to the Present* (New York: Penguin, 2012), 94-115; Helen Tilley, *Africa as a Living Laboratory: Empire, Development, and the Problem of Scientific Knowledge, 1870-1950* (Chicago: University of Chicago Press, 2011), 115-68; Paul Sutter, *Driven Wild: How the Fight against Automobiles Launched the Modern Wilderness Movement* (Seattle: University of Washington Press, 2009), 161-62.

50. On the Russian tradition of soil science that served as an important influence on and context for Vernadskii's work, see Lloyd Ackert, *Sergei Vinogradskii and the Cycle of Life: From the Thermodynamics of Life to Ecological Microbiology, 1850-1950* (Dordrecht: Springer, 2013); and Jonathan D. Oldfield and Denis J. B. Shaw, *The Development of Russian Environmental Thought: Scientific and Geographical Perspectives on the Natural Environment* (New York: Routledge, 2016), 48-77.

51. Bailes, *Science and Russian Culture*, 145, 191.

52. David Holloway, *Stalin and the Bomb: The Soviet Union and Atomic Energy, 1939–1956* (New Haven, CT: Yale University Press, 1994), 31–33; Bailes, *Science and Russian Culture*, 169.

53. On Vinogradov's relationship to Vernadskii, see Bailes, *Science and Russian Culture*, 163, 170; Boris Belitzky, "Soviet Environmentalist," *New Scientist*, January 2, 1975, 15–17; and Georgy S. Levit, "Looking at Russian Ecology through the Biosphere Theory," in *Ecology Revisited: Reflecting on Concepts, Advancing Science*, ed. Astrid Schwarz and Kurt Jax (Dordrecht: Springer, 2011), 333–47, on 341.

54. Goldschmidt quoted in C. E. Tilley, "Victor Moritz Goldschmidt: 1888–1947," *Obituary Notices of Fellows of the Royal Society* 6, no. 17 (1948): 51–66, on 55. See also Helge Kragh, "From Geochemistry to Cosmochemistry: The Origin of a Scientific Discipline, 1915–1955," in *Chemical Sciences in the 20th Century: Bridging Boundaries*, ed. Carsten Reinhardt (New York: Wiley-VCH, 2001), 160–90.

55. Assar Hadding, "Mineralienanalyse nach röntgenspektroskopischer Methode," *Zeitschift für anorganische und allgemeine Chemie* 122 (1922): 195–200. See also Mason, *Victor Moritz Goldschmidt*, 28.

56. Victor Moritz Goldschmidt, "Der Stoffwechsel der Erde," *Zeitschrift für Elektrochemie und angewandte physikalische Chemie* 28, no. 19/20 (1922): 411–21; V. M. Goldschmidt, "The Principles of Distribution of Chemical Elements in Minerals and Rocks," *Journal of the Chemical Society* [no volume or issue number] (1937): 655–73. See also Mason, *Victor Moritz Goldschmidt*, 44, 60–61

57. Bailes, *Science and Russian Culture*, 189.

58. Anne Harrington, *Reenchanted Science: Holism in German Culture from Wilhelm II to Hitler* (Princeton, NJ: Princeton University Press, 1996), 54–62.

59. Jakob von Uexküll, *Umwelt und Innenwelt der Tiere* (Berlin: Julius Springer, 1909). See also Wolf Feuerhahn, "Du milieu à l'Umwelt: Enjeux d'un changement terminologique," *Revue Philosophique de la France et de l'Étranger* 199, no. 4 (2009): 419–38.

60. Jakob von Uexküll, "Der Organismus als Staat und der Staat als Organismus," in *Der Leuchter: Weltanschauung und Lebensgestaltung*, ed. Alexander von Gliechen-Russwurm (Darmstadt: Otto Reichl Verlag, 1919), 79–110.

61. Jakob von Uexküll, *Staatsbiologie (Anatomie-Phsiologie-Pathologie des Staates)* (Berlin: Gebrüder Paetel, 1920); Jakob von Uexküll, *Theoretical Biology*, trans. D. L. Mackinnon (New York: Harcourt, Brace, 1926), 338–50. See also Harrington, *Reenchanted Science*, 54–55.

62. On the "speed" or "velocity" of life, see Vernadsky, *Biosphere*, 65–66.

63. Vito Volterra, "Fluctuations in the Abundance of a Species Considered Mathematically," *Nature* 118, no. 2973 (October 16, 1926): 558–60. See also Sharon E. Kingsland, *Modeling Nature: Episodes in the History of Population Ecology*, 2nd ed. (Chicago: University of Chicago Press, 1995), 106–7.

64. Alfred J. Lotka, *Elements of Physical Biology* (Baltimore: Williams & Wilkins, 1925). See also Ariane Tanner, *Die Mathematisierung des Lebens: Alfred James Lotka und der energetische Holismus im 20. Jahrhundert* (Tübingen: Mohr Siebeck, 2017).

65. Raymond Pearl, *The Nation's Food: A Statistical Study of a Physiological and Social Problem* (Philadelphia: W. B. Saunders, 1920); Raymond Pearl, *The Biology of Population Growth* (New York: A. A. Knopf, 1925). See also Kingsland, *Modeling Nature*, 29, 60–64;

Alison Bashford, *Global Population: History, Geopolitics, and Life on Earth* (New York: Columbia University Press, 2014), 197; and Tanner, *Die Mathematisierung des Lebens*, 84, 183.

66. G. F. Gause, "Ecology of Populations," *Quarterly Review of Biology* 7, no. 1 (1932): 27–46, quote on 44–45.

67. Douglas R. Weiner, *Models of Nature: Ecology, Conservation, and Cultural Revolution in Soviet Russia* (Bloomington: Indiana University Press, 1988), 78–82.

68. Weiner, *Models of Nature*, 82.

69. Weiner, 81.

70. G. Evelyn Hutchinson, *The Kindly Fruits of the Earth: Recollections of an Embryo Ecologist* (New Haven, CT: Yale University Press, 1979), 233; Joel B. Hagen, *An Entangled Bank: The Origins of Ecosystem Ecology* (New Brunswick, NJ: Rutgers University Press, 1992), 64–65. The spelling of George Vernadsky's last name matches the conventions for transliterating Russian names that were common at the time of his immigration to the United States.

71. G. Evelyn Hutchinson, "The Biogeochemistry of Aluminum and of Certain Related Elements," *Quarterly Review of Biology* 18, no. 1 (1943): 1–29, quote on 1.

72. Raymond L. Lindeman, "The Trophic-Dynamic Aspect of Ecology," *Ecology* 23, no. 4 (1942): 399–417. See also Robert E. Cook, "Raymond Lindeman and the Trophic-Dynamic Concept in Ecology," *Science* 198, no. 4312 (1977): 22–26.

73. Lindeman, "Trophic-Dynamic Aspect of Ecology," 400.

74. Hagen, *Entangled Bank: Origins of Ecosystem Ecology*; Frank B. Golley, *A History of the Ecosystem Concept in Ecology: More Than the Sum of the Parts* (New Haven, CT: Yale University Press, 1993); David C. Coleman, *Big Ecology: The Emergence of Ecosystem Science* (Berkeley: University of California Press, 2010).

75. Donald Worster, *Nature's Economy: A History of Ecological Ideas*, 2nd ed. (New York: Cambridge University Press, 1994), 191–204.

76. Karl Möbius, "Eine Austernbank ist eine Biocönose oder Lebensgemeinde," in *Die Auster und die Austernwirthschaft* (Berlin: Wiegandt, Hempel & Parey, 1877), 72–87, on 76. See also Lynn K. Nyhart, *Modern Nature: The Rise of the Biological Perspective in Germany* (Chicago: University of Chicago Press, 2009), 152–53.

77. Möbius, "Eine Austernbank ist eine Biocönose oder Lebensgemeinde," 76.

78. Nyhart, *Modern Nature*, 314–15.

79. Frederic Edward Clements, *Research Methods in Ecology* (Lincoln, NE: University Publishing, 1905), 199.

80. W. Vernadsky, *La Géochimie* (Paris: Félix Alcan, 1924), 52–61. See also Bailes, *Science and Russian Culture*, 186.

81. Vernadsky, *Géochimie*, 60.

82. Vernadsky, 47.

83. See Eduard Suess, *Der Antlitz der Erde*, vol. 3, part 2 (Vienna: F. Tempsky, 1909), 739–40. See also Jacques Grinevald, "Sketch for a History of the Idea of the Biosphere," in *Gaia, the Thesis, the Mechanisms and the Implications*, ed. Peter Bunyard and Edward Goldsmith (Camelford, Cornwall: Wadebridge Ecological Center, 1988), 1–34.

84. Vernadsky, *La Biosphère* (Paris: Félix Alcan, 1929), 1–5.

85. A. G. Tansley, "The Use and Abuse of Vegetational Concepts and Terms," *Ecology* 16,

no. 3 (1935): 284–307. See also Peder Anker, *Imperial Ecology: Environmental Order in the British Empire, 1895–1945* (Cambridge, MA: Harvard University Press, 2001) 118–56.

86. Tansley, "Use and Abuse of Vegetational Concepts and Terms," 296. See also Angela N. H. Creager, *Life Atomic: A History of Radioisotopes in Science and Medicine* (Chicago: University of Chicago Press, 2013), 351–53, 356–67.

87. Tansley, "Use and Abuse of Vegetational Concepts and Terms," 297; italics in the original.

88. Lindeman, "Trophic-Dynamic Aspect of Ecology," 399.

89. Lindeman, 400.

90. Nancy G. Slack, *G. Evelyn Hutchinson and the Invention of Modern Ecology* (New Haven, CT: Yale University Press, 2010), 173–74.

91. See Norbert Wiener, *Cybernetics* (New York: J. Wiley, 1948). See also Ronald R. Kline, *The Cybernetics Moment: Or Why We Call Our Age the Information Age* (Baltimore: Johns Hopkins University Press, 2015).

92. G. Evelyn Hutchinson, "Circular Causal Systems in Ecology," *Annals of the New York Academy of Sciences* 50 (1948): 221–46. See also Hagen, *Entangled Bank: Origins of Ecosystem Ecology*, 68–74; Creager, *Life Atomic*, 360; and N. Katherine Hayles, *How We Became Posthuman: Virtual Bodies in Cybernetics, Literature, and Informatics* (Chicago: University of Chicago Press, 1999), 50–83.

93. Hutchinson, "Circular Causal Systems in Ecology," 221.

94. Vernadsky, "The Biosphere and the Noösphere," 10.

95. Vernadsky, 9; italics in the original have been removed. See also Clive Hamilton and Jacques Grinevald, "Was the Anthropocene Anticipated?," *Anthropocene Review* 2, no. 1 (2015): 59–72, esp. 65–66.

96. Vernadsky, "The Biosphere and the Noösphere," 9.

97. Vernadsky, 9; italics in the original have been removed.

98. Peter J. Taylor, "Technocratic Optimism, H. T. Odum, and the Partial Transformation of Ecological Metaphor after World War II," *Journal of the History of Biology* 21, no. 2 (1988): 213–44.

99. Uexküll, *Staatsbiologie*. See also Harrington, *Reenchanted Science*, 58–62.

100. Uexküll, *Staatsbiologie*, 49–51.

101. Harrington, *Reenchanted Science*, 68–69; Florian Mildenberger and Bernd Herrmann, "Nachwort," in *Umwelt und Innenwelt der Tiere*, by Jakob Johann von Uexküll, ed. Florian Mildenberger and Bernd Herrmann (Berlin: Springer, 2014), 261–330.

102. See William Vogt, *Road to Survival* (New York: W. Sloane Associates, 1948). See also Cushman, *Guano and the Pacific World*, 243–81; and Thomas Robertson, "Total War and the Total Environment: Fairfield Osborn, William Vogt, and the Birth of Global Ecology," *Environmental History* 17, no. 2 (2012): 336–64.

103. Bailes, *Science and Russian Culture*, 170–74; Holloway, *Stalin and the Bomb*, 59–63.

104. Holloway, *Stalin and the Bomb*, 100–115; Kendall E. Bailes, "Soviet Science in the Stalin Period: The Case of V. I. Vernadskii and His Scientific School, 1928–1945," *Slavic Review* 45, no. 1 (1986): 20–37, on 28–32.

105. J. K. Gustafson, "Uranium Resources," *Scientific Monthly* 69, no. 2 (1949): 115–20, on

119; Richard G. Hewlett and Oscar E. Anderson Jr., *The New World, 1939–1946: A History of the United States Atomic Energy Commission*, vol. 1 (University Park: Pennsylvania State University Press, 1962). See also Jonathan E. Helmreich, *Gathering Rare Ores: The Diplomacy of Uranium Acquisition, 1943–1954* (Princeton, NJ: Princeton University Press, 1986).

106. Glenn T. Seaborg, "Plutonium and Other Transuranium Elements," *Chemical and Engineering News* 25, no. 6 (1947): 358–97, on 359–60.

107. United States Department of Energy, *Plutonium: The First 50 Years* (Washington, DC: US Department of Energy, 1996), 3.

108. Slack, *G. Evelyn Hutchinson and the Invention of Modern Ecology*, 159–70; Creager, *Life Atomic*, 357–63.

109. Creager, *Life Atomic*, 137–41; Jacob Hamblin, "Exorcising Ghosts in the Age of Automation: United Nations Experts and Atoms for Peace," *Technology and Culture* 47, no. 4 (2006): 734–56; John Krige, "Atoms for Peace, Scientific Internationalism, and Scientific Intelligence," in *Global Power Knowledge: Science and Technology in International Affairs*, vol. 21 of *Osiris*, ed. John Krige and Kai-Henrik Barth (Chicago: University of Chicago Press, 2006), 161–81.

110. Matthew Evangelista, *Unarmed Forces: The Transnational Movement to End the Cold War* (Ithaca, NY: Cornell University Press, 1999), 51–52; Holloway, *Stalin and the Bomb*, 337–39; Toshihiro Higuchi, "Epistemic Frictions: Radioactive Fallout, Health Risk Assessments, and the Eisenhower Administration's Nuclear-Test Ban Policy, 1954–1958," *International Relations of the Asia-Pacific* 18, no. 1 (2018): 99–124, on 115.

111. A. P. Vinogradov, "Prospects for the Pugwash Movement," *Bulletin of the Atomic Scientists* 15, no. 9 (1959): 376–78, quote on 377.

112. Taylor, "Technocratic Optimism, Odum, and Ecological Metaphor."

113. Odum's dissertation was titled "The Biogeochemistry of Strontium: With Discussion on the Ecological Integration of Elements"; some of its findings were published in H. T. Odum, "Stability of the World Strontium Cycle," *Science* 114, no. 2964 (1951): 407–11. See also Karin E. Limburg, "The Biogeochemistry of Strontium: A Review of H. T. Odum's Contributions," *Ecological Modelling* 178 (2004): 31–33.

114. Howard T. Odum and Eugene P. Odum, "Trophic Structure and Productivity of a Windward Coral Reef Community on Eniwetok Atoll," *Ecological Monographs* 25, no. 3 (1955): 291–320, on 301. See also Laura J. Martin, "Proving Grounds: Ecological Fieldwork in the Pacific and the Materialization of Ecosystems," *Environmental History* 23, no. 3 (2018): 567–92. The spelling of the atoll's name has changed from Eniwetok to Enewetak in the years since the Odums' study.

115. E. P. Odum, R. P. Martin, and B. C. Loughman, "Scanning Systems for the Rapid Determination of Radioactivity in Ecological Materials," *Ecology* 43, no. 1 (1962): 171–73, quote on 171. See also Creager, *Life Atomic*, 386–88; and Chunglin Kwa, "Radiation Ecology, Systems Ecology and the Management of the Environment," in *Science and Nature: Essays in the History of the Environmental Sciences*, ed. Michael Shortland, BSHS Monographs 8 (British Society for the History of Science, 1993), 213–49.

116. Paul N. Edwards, *The Closed World: Computers and the Politics of Discourse in Cold War America* (Cambridge, MA: MIT Press, 1996), 43–74; Kristine C. Harper, *Weather by the Numbers: The Genesis of Modern Meteorology* (Cambridge, MA: MIT Press, 2008), 96–104.

117. G. Ledyard Stebbins, "International Biological Program," *Science* 137, no. 3532 (1962): 768–70.

118. Frederick E. Smith, "The International Biological Program and the Science of Ecology," *Proceedings of the National Academy of Sciences of the United States of America* 60, no. 1 (1968): 5–11, on 6.

119. Christophe Bonneuil and Jean-Baptiste Fressoz, *The Shock of the Anthropocene: The Earth, History, and Us*, trans. David Fernbach (New York: Verso, 2015), 122–47.

120. Vernadsky, "The Biosphere and the Noösphere," 5.

CHAPTER 5

1. Wayne C. Hall, *Amino Triazole: A New Abscission Chemical and Growth Inhibitor* (College Station: Texas Agricultural Experiment Station, 1954), 5; R. Milton Carleton, "Latest Garden News," *Better Homes and Gardens* 34, no. 10 (1956): 13. See also Robert L. Zimdahl, *A History of Weed Science in the United States* (London: Elsevier, 2010), 96–97.

2. "Some of Cranberry Crop Tainted by a Weed-Killer, U.S. Warns," *New York Times*, November 10, 1959. See also M. R. Janzen, "The Cranberry Scare of 1959: The Beginning of the End of the Delaney Clause" (PhD diss., Texas A&M University, 2010).

3. J. Richard Beattie, "Mass. Cranberry Station and Field Notes," *Cranberries* 4, no. 7 (November 1959): 3–4, quote on 3; Betty Carrollton, "Cranberries Taken Off Shelves Here," *Atlanta Constitution*, November 11, 1959.

4. "Producers Rip Cranberry Report," *Chicago Daily Tribune*, November 11, 1959; "Unexpected Marketing Crisis Rocks the Entire Industry," *Cranberries* 4, no. 7 (November 1959): 11–13.

5. "Nixon Comments on Cranberries," *Washington Post*, November 15, 1999.

6. "Cranberry Growers to Sue U.S.," *Washington Post*, December 22, 1959.

7. "United States to Pay Indemnity for Cranberry Losses," *Science* 131, no. 3406 (1960), 1033–34; "Pre-Easter Sales Mark Turning Point in the Industry," *Cranberries* 25, no. 1 (May 1960): 18.

8. US Department of Agriculture, Crop Reporting Board, *Fruits, Non-Citrus, by States, 1959–64: Production, Use, Value* (Washington, DC: USDA, Statistical Reporting Service, 1967), 36. See also "Cancer Scare Forgotten, Cranberry Farms Thrive," *New York Times*, November 18, 1979.

9. Janzen, "Cranberry Scare of 1959," 101–40.

10. Rachel Carson, *Silent Spring* (1962; Boston: Houghton Mifflin, 2002), 8.

11. John Maynard Keynes, *The General Theory of Employment, Interest and Money* (New York: Harcourt, Brace, 1936), 372–84.

12. Meg Jacobs, *Pocketbook Politics: Economic Citizenship in Twentieth-Century America* (Princeton, NJ: Princeton University Press, 2007), 136–75; Lizabeth Cohen, *A Consumers' Republic: The Politics of Mass Consumption in Postwar America* (New York: Vintage Books, 2004), 28–31.

13. Jacobs, *Pocketbook Politics*, 179–220; Cohen, *Consumers' Republic*, 62–109.

14. Cohen, *Consumers' Republic*, 301.

15. Ruth Schwartz Cowan, *More Work for Mother: The Ironies of Household Technology from the Open Hearth to the Microwave* (New York: Basic Books, 1983), 315–16.

16. Patricia Kirkham, *Charles and Ray Eames: Designers of the Twentieth Century* (Cambridge: MIT Press, 1998), 201–31.

17. Jeffrey L. Meikle, *American Plastic: A Cultural History* (New Brunswick, NJ: Rutgers University Press, 1995), 189–91.

18. Robert A. Beauregard, *When America Became Suburban* (Minneapolis: University of Minnesota Press, 2006), 33–39; Kenneth T. Jackson, *Crabgrass Frontier: The Suburbanization of the United States* (New York: Oxford University Press, 1985) 289–90; Thomas Sugrue, *The Origins of the Urban Crisis: Race and Inequality in Postwar Detroit* (Princeton, NJ: Princeton University Press, 1996), 245–46.

19. Adam W. Rome, *The Bulldozer in the Countryside: Suburban Sprawl and the Rise of American Environmentalism* (New York: Cambridge University Press, 2001), 45–86.

20. Beauregard, *When America Became Suburban*, 94.

21. Susanne Freidberg, *Fresh: A Perishable History* (Cambridge, MA: Belknap Press of Harvard University Press, 2009), 169; Shane Hamilton, *Trucking Country: The Road to America's Wal-Mart Economy* (Princeton, NJ: Princeton University Press, 2008), 99–134.

22. William Cronon, *Nature's Metropolis: Chicago and the Great West* (New York: W. W. Norton, 1991), 255–59; Sara B. Pritchard and Thomas Zeller, "The Nature of Industrialization," in *The Illusory Boundary: Environment and Technology in History*, ed. Martin Reuss and Stephen H. Cutcliffe (Charlottesville: University of Virginia Press, 2010), 69–101.

23. Christopher Sellers, *Crabgrass Crucible: Suburban Nature and the Rise of Environmentalism in Twentieth Century America* (Chapel Hill: University of North Carolina Press, 2012), 80–81.

24. For a comprehensive review of studies on aminotriazole's carcinogenicity, see IARC, "Amitrole," in *Some Thyrotropic Agents*, vol. 79 of IARC Monographs on the Evaluation of Carcinogenic Risk of Chemicals to Humans (Lyon, France: International Agency for Research on Cancer, 2001), 381–410.

25. Christopher C. Sellers, *Hazards of the Job: From Industrial Disease to Environmental Health Science* (Chapel Hill: University of North Carolina Press, 1997), 206–7.

26. Sellers, *Hazards of the Job*, 193.

27. R. A. Kehoe, F. Thamann, and J. Cholak, "On the Normal Absorption and Excretion of Lead: I. Lead Absorption and Excretion in Primitive Life," *Journal of Industrial Hygiene* 15 (1933): 257–72; R. A. Kehoe, F. Thamann, and J. Cholak, "On the Normal Absorption and Excretion of Lead: II. Lead Absorption and Lead Excretion in Modern American Life," *Journal of Industrial Hygiene* 15 (1933): 273–79. See also Sellers, *Hazards of the Job*, 216–18.

28. Robert A. Kehoe, "The Modern Environment," *Public Health Reports* 73, no. 12 (1958): 1074–75. See also Sellers, *Hazards of the Job*, 224.

29. Sellers, *Crabgrass Crucible*, 259–60.

30. Richard J. Block, Emmett L. Durrum, and Gunter Zweig, *A Manual of Paper Chromatography and Paper Electrophoresis* (New York: Academic Press, 1955).

31. Jean White, "FDA Chemists' Work Is a Bowl of Berries," *Washington Post*, November 13, 1959; F. D. Aldrich and S. R. McLane Jr., "A Paper Chromatographic Method for the Detection of 3-Amino-1,2,4-Triazole in Plant Tissues," *Plant Physiology* 32, no. 2 (1957): 153–54; Paul A. Mills, "Detection and Semiquantitative Estimation of Chlorinated Organic Pesticide Residues

in Foods by Paper Chromatography," *Journal of the Association of Official Analytical Chemists* 42 (1959): 734–40.

32. See, e.g., P. A. Neal, T. R. Sweeney, S. S. Spicer, and W. F. von Oettingen, "The Excretion of DDT (2,2-Bis-(P-Chlorophenyl)-1,1,1-Trichloroethane) in Man, Together with Clinical Observations," *Public Health Reports* 61, no. 12 (1946): 403–9.

33. See, e.g., William R. Hill and C. Robert Damiani, "Death Following Exposure to DDT—Report of a Case," *New England Journal of Medicine* 235 (1946): 897–99.

34. Frederick Rowe Davis, *Banned: A History of Pesticides and the Science of Toxicology* (New Haven, CT: Yale University Press, 2014), 26–28.

35. Jeffrey M. Paull, "The Origin and Basis of Threshold Limit Values," *American Journal of Industrial Medicine* 5, no. 3 (1984): 227–38; Sellers, *Hazards of the Job*, 175–76.

36. T. H. Jukes and C. B. Shaffer, "Antithyroid Effects of Aminotriazole," *Science* 132, no. 3422 (1960): 296–97. See also Janzen, "Cranberry Scare of 1959," 65–68.

37. "Chemists' Findings," *Wall Street Journal*, November 12, 1959.

38. "Chemists' Findings," *Wall Street Journal*.

39. Thomas H. Jukes, "Some Biological Effects of Aminotriazole," *Economic Botany* 17, no. 3 (1963): 238–40, quote on 240. See also Robert N. Procter, *Cancer Wars: How Politics Shapes What We Know and Don't Know about Cancer* (New York: Basic Books, 1995), 171–73; and Elena Conis, "Debating the Health Effects of DDT: Thomas Jukes, Charles Wurster, and the Fate of an Environmental Pollutant," *Public Health Reports* 125, no. 2 (2010): 337–42.

40. See Bruce N. Ames, William E. Durston, Edith Yamasaki, and Frank D. Lee, "Carcinogens Are Mutagens: A Simple Test System Combining Liver Homogenates for Activation and Bacteria for Detection," *Proceedings of the National Academy of Sciences of the United States of America* 70, no. 8 (1973): 2281–85.

41. Bruce N. Ames, H. O. Kammen, and Edith Yamasaki, "Hair Dyes Are Mutagenic: Identification of a Variety of Mutagenic Ingredients," *Proceedings of the National Academy of Sciences of the United States of America* 72, no. 6 (1975): 2423–27; Arlene Blum and Bruce N. Ames, "Flame-Retardant Additives as Possible Cancer Hazards," *Science* 195, no. 4273 (1977): 17–23.

42. Bruce N. Ames, "Dietary Carcinogens and Anticarcinogens," *Science* 221, no. 4617 (1983): 1256–64, quote on 1256. See also Angela N. H. Creager, "The Political Life of Muta gens: A History of the Ames Test," in *Powerless Science? Science and Politics in a Toxic World*, ed. Soraya Boudia and Nathalie Jas (New York: Berghahn, 2014), 46–64.

43. Clarence Cottam and Elmer Higgins, "DDT and Its Effect on Fish and Wildlife," *Journal of Economic Entomology* 39 (1946): 44–52; Tracy I. Storer, "DDT and Wildlife," *Journal of Wildlife Management* 10, no. 3 (1946): 181–83. See also Thomas Dunlap, *DDT: Scientists, Citizens, and Public Policy* (Princeton, NJ: Princeton University Press, 1981), 76.

44. Food, Drug, and Cosmetic Act, Pub. L. No. 75-717, 53 Stat. 1040 (1938). See also Gwen Kay, "Healthy Public Relations: The FDA's 1930s Legislative Campaign," *Bulletin of the History of Medicine* 75, no. 3, (2001): 446–87. The most important of these scandals was the death of more than one hundred victims from an improperly prepared medicine called Elixir Sulfanilamide in 1937.

45. Pub. L. No. 80-104, 61 Stat. 163 (1947). See also Christopher Bosso, *Pesticides and Politics: The Life Cycle of a Public Issue* (Pittsburgh: University of Pittsburgh Press, 1997), 10–11;

and Courtney I. P. Thomas, *In Food We Trust: The Politics of Purity in American Food* (Lincoln: University of Nebraska Press, 2014), 41–57.

46. Davis, *Banned: A History of Pesticides*, 121–49.

47. See, e.g., Wilhelm C. Hueper, "Carcinogens and Carcinogenesis," *American Journal of Medicine* 8, no. 3 (1950): 355–71. See also Christopher Sellers, "Discovering Environmental Cancer: Wilhelm Hueper, Post–World War II Epidemiology, and the Vanishing Clinician's Eye," *American Journal of Public Health* 87, no. 11 (1997): 1824–35.

48. Lear, *Rachel Carson*, 358–60; Lewis Herber [Murray Bookchin], *Our Synthetic Environment* (New York: Knopf, 1962), xv.

49. Pub. L. No. 85-929, 72 Stat. 1784 (1958). See also Nancy Langston, *Toxic Bodies: Hormone Disruptors and the Legacy of DES* (New Haven, CT: Yale University Press, 2010), 80–82.

50. John A. Osmundsen, "Food Trade Waits Impact of U.S. Law on Additives," *New York Times*, February 23, 1959.

51. Sellers, *Crabgrass Crucible*, 126–28; see also 105–37.

52. Robert E. Smolker, "The Environmental Defense Fund," *Biological Conservation* 1, no. 1 (October 1968): 69–70. On these activists' adoption of the term *environment*, see Sellers, *Crabgrass Crucible*, 245–46.

53. For "biocides," see Carson, *Silent Spring*, 8. For references to aminotriazole (or "amitrole," as Carson calls it) and the cranberry scare, see Carson, 36, 182, 225–26. See also Lear, *Rachel Carson*, 358–60; and William Souder, *On a Farther Shore: The Life and Legacy of Rachel Carson* (New York: Crown Publishers, 2012), 305.

54. *International Coordination in Environmental Hazards (Pesticides), Hearings before the Subcommittee on Reorganization and International Organizations of the Committee on Government Operations*, US Senate, 88th Congress, 1st Session, part 1 (Washington, DC: US Government Printing Office, 1964), 7.

55. *International Coordination in Environmental Hazards (Pesticides), Hearings before the Subcommittee on Reorganization and International Organizations of the Committee on Government Operations*, US Senate, 88th Congress, 1st Session, part 3 (Washington, DC: US Government Printing Office, 1964), 697.

56. Lynton K. Caldwell, "Environment: A New Focus for Public Policy?," *Public Administration Review* 23, no. 3 (1963): 132–39. See also Wendy Read Wertz, *Lynton Keith Caldwell: An Environmental Visionary and the National Environmental Policy Act* (Bloomington: Indiana University Press, 2014), 82–87.

57. Federal Environmental Pesticide Control Act, Pub. L. No. 92-516, 86 Stat. 973 (1972); Toxic Substances Control Act, Pub. L. No. 94-469, 90 Stat. 2003 (1976). See also Davis, *Banned: A History of Pesticides*, 192; and Sarah A. Vogel and Jody A. Roberts, "Why the Toxic Substances Control Act Needs an Overhaul, and How to Strengthen Oversight of Chemicals in the Interim," *Health Affairs* 30, no. 5 (2011): 898–905.

58. John Wargo, *Our Children's Toxic Legacy: How Science and Law Fail to Protect Us from Pesticides* (New Haven, CT: Yale University Press, 1996), 89.

59. Charles E. Rosenberg, "Pathologies of Progress: The Idea of Civilization as Risk," *Bulletin of the History of Medicine* 72, no. 4 (1998): 714–30.

60. See Julian Huxley, *Evolution: The Modern Synthesis* (New York: Harper and Brothers, 1942). See also Vassiliki Betty Smocovitis, *Unifying Biology: The Evolutionary Synthesis and Evolutionary Biology* (Princeton, NJ: Princeton University Press, 1996), 162–63.

61. Theodosius Dobzhansky, *Genetics and the Origin of Species* (New York: Columbia University Press, 1937), 120. See also William B. Provine, "The Origins of Dobzhansky's *Genetics and the Origin of Species*," in *The Evolution of Theodosius Dobzhansky: Essays on His Life and Thought in Russia and America*, ed. Mark B. Adams (Princeton, NJ: Princeton University Press, 1994), 99–114.

62. On Carson's exploration of evolution in her earlier work *Under the Sea Wind: A Naturalist's Picture of Ocean Life* (New York: Simon and Schuster, 1941), see Mark Hamilton Lytle, *The Gentle Subversive: Rachel Carson, "Silent Spring", and the Rise of the Environmental Movement* (New York: Oxford University Press, 2007), 87–88.

63. Lear, *Rachel Carson*, 292. The book was Julian Huxley, *Evolution in Action* (New York: Harper, 1953).

64. The phrase is from a letter written by Carson to her friend Dorothy Freeman on February 1, 1958, in *Always, Rachel: The Letters of Rachel Carson and Dorothy Freeman, 1952–1964*, ed. Martha Freeman (Boston: Beacon, 1995), 248. See also Lisa Sideris, "The Ecological Body," *Soundings: An Interdisciplinary Journal* 85, no. 1/2 (2002): 107–20, on 108.

65. Carson, *Silent Spring*, 7.

66. Smocovitis, *Unifying Biology*, 145–48.

67. Carson, *Silent Spring*, 208–16, quote on 208. See also Ralph H. Lutts, "Chemical Fallout: Rachel Carson's *Silent Spring*, Radioactive Fallout, and the Environmental Movement," *Environmental Review* 9, no. 3 (1985): 210–25; and Scott Frickel, *Chemical Consequences: Environmental Mutagens, Scientist Activism, and the Rise of Genetic Toxicology* (New Brunswick: Rutgers University Press, 2004), 9–10.

68. Carson, *Silent Spring*, 8.

69. Rachel Carson, "The Pollution of Our Environment," in *Lost Woods: The Discovered Writing of Rachel Carson*, ed. Linda Lear (Boston: Beacon Press, 1998), 227–45, quote on 245. On Carson's diagnosis and treatment for breast cancer, see Robert Aronowitz, *Unnatural History: Breast Cancer and American Society* (New York: Cambridge University Press, 2007), 183–209.

70. "Statement on the Nature of Race and Race Differences, Paris, June 1951," in *Four Statements on the Race Question*, by United Nations Educational, Scientific, and Cultural Organization (Paris: UNESCO, 1969), 36–43.

71. Herber [Bookchin], *Our Synthetic Environment*. See also Janet Biehl, *Ecology or Catastrophe: The Life of Murray Bookchin* (New York: Oxford University Press, 2015), 85–86.

72. Herber [Bookchin], *Our Synthetic Environment*, 27–28.

73. Herber [Bookchin], 28.

74. See Carson, *Silent Spring*, 1–3; and Christine Oravec, "An Inventional Archaeology of 'A Fable for Tomorrow,'" in *And No Birds Sing: Rhetorical Analyses of Rachel Carson's "Silent Spring"*, ed. Craig Waddell (Carbondale: Southern Illinois University Press, 2000), 42–59.

75. Carson, *Silent Spring*, 278.

76. René Dubos, *Man Adapting* (New Haven, CT: Yale University Press, 1965), 278–79. See

also Carol L. Moberg and Zanvil A. Cohn, "René Jules Dubos," *Scientific American* 264, no. 5 (1991): 66–75, on 74.

77. René Dubos, "Man versus Environment," *Industrial Medicine and Surgery* 30 (1961): 369–73, on 373.

78. René Dubos, "The Despairing Optimist," *American Scholar* 40, no. 1 (1970/71): 16–20, on 20.

79. On his opposition to the Delaney Clause, see Dubos, "Man versus Environment," 372; and Dubos, *Man Adapting*, 385–87.

80. "The Despairing Optimist" was the title of a regular column that Dubos wrote for the *American Scholar* beginning in 1970. For the first installment, see *American Scholar* 40, no. 1 (1970/71): 16–20.

81. Dubos, *Man Adapting*.

82. Robert D. Bullard, *Dumping in Dixie: Race, Class, and Environmental Quality* (Boulder, CO: Westview Press, 1990).

83. Naomi Oreskes and Erik M. Conway, *Merchants of Doubt: How a Handful of Scientists Obscured the Truth on Issues from Tobacco Smoke to Global Warming* (New York: Bloomsbury Press, 2010), 134.

84. Steve Fraser and Gary Gerstle, eds., *The Rise and Fall of the New Deal Order, 1930–1980* (Princeton, NJ: Princeton University Press, 1989), 276–78.

85. Hal K. Rothman, *Saving the Planet: The American Response to the Environment in the Twentieth Century* (Chicago: Ivan R. Dee, 2000), 204; Gary S. Cross, *An All-Consuming Century: Why Commercialism Won in Modern America* (New York: Columbia University Press, 2000) 159; Cohen, *Consumers' Republic*, 387.

86. Davis, *Banned: A History of Pesticides*, 189–91.

87. Claude Reed Jr., "The PCB Problem," *New Journal and Guide*, December 29, 1982.

88. Quoted in Dale Russakoff, "As in the '60s, Protesters Rally; but This Time the Foe Is PCB," *Washington Post*, October 11, 1982.

89. Commission for Racial Justice, *Toxic Wastes and Race in the United States: A National Report on the Racial and Socio-economic Characteristics of Communities with Hazardous Waste Sites* (New York: United Church of Christ, 1987); "Civil Rights Director Charges 'Environmental Racism,'" *Afro-American*, May 9, 1987. See also Eileen McGurty, *Transforming Environmentalism: PCBs, Warren County and the Origins of Environmental Justice* (New Brunswick: Rutgers University Press, 2007).

90. Robert Bullard, *Unequal Protection: Environmental Justice and Communities of Color* (San Francisco: Sierra Club Books, 1996), 200–202.

91. Phil Brown, "Popular Epidemiology and Toxic Waste Contamination: Lay and Professional Ways of Knowing," *Journal of Health and Social Behavior* 33, no. 3 (1992): 267–81; Michelle Murphy, *Sick Building Syndrome and the Problem of Uncertainty: Environmental Politics, Technoscience, and Women Workers* (Durham, NC: Duke University Press, 2006), 81–110.

92. Lois Marie Gibbs, *Love Canal: My Story* (Albany: State University of New York Press, 1982), 66–69; Beverly Paigen, "Controversy at Love Canal," *Hastings Center Report* 12, no. 3 (1982): 29–37; Lois Gibbs, "Risk Assessments from a Community Perspective," *Environmental Impact Assessment Review* 14, no. 5–6 (1994): 327–35. See also Elizabeth D. Blum, *Love Canal*

Revisited: Race, Class, and Gender in Environmental Activism (Lawrence: University Press of Kansas, 2008).

93. Commission for Racial Justice, *Toxic Wastes and Race in the United States*, 15.

94. Charles Lee, ed., *Proceedings: The First National People of Color Environmental Leadership Summit* (New York: United Church of Christ Commission for Racial Justice, 1992).

95. Samuel P. Hays in collaboration with Barbara D. Hays, *Beauty, Health, and Permanence: Environmental Politics in the United States, 1955–1985* (New York: Cambridge University Press, 1987), 458–90; Robert Cameron Mitchell, "Twenty Years of Environmental Mobilization: Trends among National Environmental Organizations," *Society & Natural Resources* 4, no. 3 (1991): 219–39.

96. Dorceta E. Taylor, "Environmentalism and the Politics of Inclusion," in *Confronting Environmental Racism: Voices from the Grassroots*, ed. Robert D. Bullard (Boston, MA: South End Press, 1993), 53–61.

97. Will Collette, "Institutions: Citizen's Clearinghouse for Hazardous Wastes," *Environment* 29, no. 9 (1987): 44–45, quote on 44. See also Newman, *Love Canal*, 212–17.

98. Melissa A. Checker, "'It's in the Air': Redefining the Environment as a New Metaphor for Old Social Justice Struggles," *Human Organization* 61, no. 1 (2002): 94–106; Patrick Novotny, *Where We Live, Work and Play: The Environmental Justice Movement and the Struggle for a New Environmentalism* (Westport, CT: Praeger, 2000), 77–78; Giovanna Di Chiro, "Nature as Community: The Convergence of Environmental and Social Justice," in *Uncommon Ground: Rethinking the Human Place in Nature*, ed. William Cronon (New York: W. W. Norton, 1996), 298–320.

99. Stan Benjamin, "Ecology Movement Flexes Legal Muscle," *Washington Post*, December 29, 1971. See also Hays, *Beauty, Health, and Permanence*, 458–90.

100. Laura Pulido, *Environmentalism and Economic Justice: Two Chicano Struggles in the Southwest* (Tucson: University of Arizona Press, 1996), 3–30.

101. SouthWest Organizing Project, "Letter to Big Ten Environmental Groups," March 16, 1990, in *Environmental Justice in Postwar America: A Documentary Reader*, ed. Christopher W. Wells (Seattle: University of Washington Press, 2018), 164–68.

102. Carson, *Silent Spring*, 5. See also Lear, *Rachel Carson*, 54–80.

103. Cf. Dunlap, *DDT: Scientists, Citizens, Public Policy*; and Thomas R. Dunlap, *Saving America's Wildlife: Ecology and the American Mind, 1850–1990* (Princeton, NJ: Princeton University Press, 1988).

104. Adam Rome, *The Genius of Earth Day: How a 1970 Teach-In Unexpectedly Made the First Green Generation* (New York: Hill and Wang, 2013), 9–10.

CHAPTER 6

1. See J. T. Houghton, G. J. Jenkins, and J. J. Ephraums, eds., *Climate Change: The IPCC Scientific Assessment* (Cambridge: Cambridge University Press, 1990).

2. IPCC Working Group I, "Policymakers Summary," in Houghton, Jenkins, and Ephraums, *Climate Change*, vii–xxxiv, on xii, xvi.

3. IPCC Working Group I, "Policymakers Summary," xxxii.

4. United Nations Framework Convention on Climate Change, S. Treaty Doc. No. 102-38, 1771 U.N.T.S. 107 (May 9, 1992). See also Joshua P. Howe, *Behind the Curve: Science and the Politics of Global Warming* (Seattle: University of Washington Press, 2014), 180–85.

5. *UN Framework Convention on Climate Change*, 1.

6. *UN Framework Convention on Climate Change*, 5.

7. Matthias Heymann, "The Evolution of Climate Ideas and Knowledge," *WIREs Climate Change* 1 (2010): 581–97.

8. John Cook, Dana Nuccitelli, Sarah A. Green, Mark Richardson, Bärbel Winkler, Rob Painting, Robert Way, Peter Jacobs, and Andrew Skuce, "Quantifying the Consensus on Anthropogenic Global Warming in the Scientific Literature," *Environmental Research Letters* 8, no. 2 (2013), https://doi.org/10.1088/1748-9326/8/2/024024.

9. World Meteorological Association, *Annual Report for 1951* (Geneva: World Meteorological Organization, 1952), https://library.wmo.int/pmb_ged/wmo_7_en.pdf.

10. Paul N. Edwards, "Meteorology as Infrastructural Globalism," in *Global Power Knowledge: Science and Technology in International Affairs*, vol. 21 of *Osiris*, ed. John Krige and Kai-Henrik Barth (Chicago: University of Chicago Press, 2006), 229–50.

11. Ronald E. Doel, "Constituting the Postwar Earth Sciences: The Military's Influence on the Environmental Sciences in the USA after 1945," *Social Studies of Science* 33, no. 5 (2003): 635–66; Naomi Oreskes, "Changing the Mission: From the Cold War to Climate Change," in *Science and Technology in the Global Cold War*, ed. Naomi Oreskes and John Krige (Cambridge, MA: MIT Press, 2014), 141–88.

12. Chandra Mukerji, *A Fragile Power: Scientists and the State* (Princeton, NJ: Princeton University Press, 1989); Naomi Oreskes, "A Context of Motivation: US Navy Oceanographic Research and the Discovery of Sea-Floor Hydrothermal Vents," *Social Studies of Science* 33, no. 5 (2003): 697–742; Allan A. Needell, *Science, Cold War and the American State: Lloyd V. Berkner and the Balance of Professional Ideals* (Amsterdam: Harwood Academic Publishers, 2000).

13. Jacob Darwin Hamblin, *Arming Mother Nature: The Birth of Catastrophic Environmentalism* (New York: Oxford University Press, 2013).

14. Joint Organizing Committee of the Global Atmospheric Research Programme, *An Introduction to GARP*, GARP Publication Series No. 1 (Geneva: World Meteorological Organization; Paris: International Council of Scientific Unions, 1969). See also Paul N. Edwards, *A Vast Machine: Computer Models, Climate Data, and the Politics of Global Warming* (Cambridge, MA: MIT Press, 2010), 242–50.

15. Joint Organizing Committee of the Global Atmospheric Research Programme, *The First GARP Global Experiment: Objectives and Plans*, GARP Publications Series, no. 11 (Geneva: World Meteorological Organization; Paris: International Council of Scientific Unions, 1973), 14.

16. See, e.g., Joint Organizing Committee of the Global Atmospheric Research Programme, *The GARP Programme on Numerical Experimentation*, GARP Publications Series, no. 7 (Geneva: World Meteorological Organization; Paris: International Council of Scientific Unions, 1971).

17. Perrin Selcer, *The Postwar Origins of the Global Environment: How the United Nations Built Spaceship Earth* (New York: Columbia University Press, 2018), 236–37.

18. UN General Assembly, Resolution 2997, "Institutional and Financial Arrangements for International Environmental Cooperation," in *Report of the United Nations Conference on En-*

vironment and Development, A/RES/27/2997 (December 15, 1972), 43–45, http://www.un.org /en/ga/search/view_doc.asp?symbol=a/res/2997(XXVII).

19. See World Commission on Environment and Development, *Our Common Future* (New York: Oxford University Press, 1987).

20. UN General Assembly, "Rio Declaration on Environment and Development," in *Report of the United Nations Conference on Environment and Development*, A/CONF.151/26 (August 12, 1992), http://www.unesco.org/education/pdf/RIO_E.PDF. See also W. M. Adams, *Green Development: Environment and Sustainability in the Third World*, 2nd ed. (London: Routledge, 2001).

21. Will Steffen, Paul J. Crutzen, and John R. McNeill, "The Anthropocene: Are Humans Now Overwhelming the Great Forces of Nature?," *Ambio* 36, no. 8 (2007): 614–21; J. R. McNeill and Peter Engelke, *The Great Acceleration: An Environmental History of the Anthropocene since 1945* (Cambridge, MA: Belknap Press of Harvard University Press, 2014).

22. Georges-Louis Leclerc, Comte de Buffon, *Époques de la nature*, vol. 2 (Paris: L'Imprimerie Royale, 1780), 197–98. See also Fabien Locher and Jean-Baptiste Fressoz, "Modernity's Frail Climate: A Climate History of Environmental Reflexivity," *Critical Inquiry* 38, no. 3 (2012): 579–98.

23. Joseph Fourier, "Remarques générales sur les températures du globe terrestre et des espaces planétaires," *Annales de Chimie et de Physique* 27 (1824): 136–67; Eunice Foote, "Circumstances Affecting the Heat of the Sun's Rays," *American Journal of Science and Arts* 22 (1856): 382–83; John Tyndall, "Note on the Transmission of Radiant Heat through Gaseous Bodies," *Proceedings of the Royal Society of London* 10 (1859): 37–39; Svante Arrhenius, "On the Influence of Carbonic Acid in the Air upon the Temperature of the Ground," *Philosophical Magazine and Journal of Science*, series 5, vol. 41 (1896): 237–76; G. S. Callendar, "The Artificial Production of Carbon Dioxide and Its Influence on Temperature," *Quarterly Journal of the Royal Meteorological Society* 64 (1938): 223–40. See also Spencer R. Weart, *The Discovery of Global Warming* (Cambridge, MA: Harvard University Press, 2008); James Rodger Fleming, *Historical Perspectives on Climate Change* (New York: Oxford University Press, 1998), 55–82; James R. Fleming, *The Callendar Effect: The Life and Work of Guy Stewart Callendar (1898– 1964)* (Boston: American Meteorological Society, 2007); and Roland Jackson, "Eunice Foote, John Tyndall and a Question of Priority," *Notes and Records: The Royal Society Journal of the History of Science*, February 13, 2019, https://doi.org/10.1098/rsnr.2018.0066.

24. Fleming, *Historical Perspectives on Climate Change*, 107–28.

25. Callendar, "Artificial Production of Carbon Dioxide," 236.

26. Roger Revelle and Hans E. Suess, "Carbon Dioxide Exchange between Atmosphere and Ocean and the Question of an Increase of Atmospheric Carbon Dioxide during the Past Decades," *Tellus* 9 (1957): 18–27. See also Fleming, *Historical Perspectives on Climate Change*, 124–26.

27. Revelle and Suess, "Carbon Dioxide Exchange between Atmosphere and Ocean," 26.

28. Gilbert N. Plass, "Carbon Dioxide and the Climate," *American Scientist* 44, no. 3 (1956): 302–16. See also James Rodger Fleming, "Gilbert N. Plass: Climate Science in Perspective," *American Scientist* 98, no. 1 (2010): 60–61.

29. Kristine C. Harper, *Weather by the Numbers: The Genesis of Modern Meteorology* (Cambridge, MA: MIT Press, 2008); Edwards, *Vast Machine*, 111–37.

30. Norman A. Phillips, "The General Circulation of the Atmosphere: A Numerical Exper-

iment," *Quarterly Journal of the Royal Meteorological Society* 82, no. 352 (1956): 123–64. See also Edwards, *Vast Machine*, 150–53.

31. See, e.g., Joseph Smagorinsky, Syukuro Manabe, and J. Leith Holloway, "Numerical Results from a Nine-Level General Circulation Model of the Atmosphere," *Monthly Weather Review* 93 (December 1965): 727–68.

32. Edwards, *Vast Machine*, 153–62.

33. Jack C. Pales and Charles D. Keeling, "The Concentration of Atmospheric Carbon Dioxide in Hawaii," *Journal of Geophysical Research* 70 (1965): 6053–76. See also Howe, *Behind the Curve*, 16–43.

34. Charles D. Keeling, "Is Carbon Dioxide from Fossil Fuel Changing Man's Environment?," *Proceedings of the American Philosophical Society* 114, no. 1 (1970): 10–17. See also Howe, *Behind the Curve*, 3–6.

35. Alessandro Antonello and Mark Carey, "Ice Cores and the Temporalities of the Global Environment," *Environmental Humanities* 9, no. 2 (2017): 181–203.

36. See Syukuro Manabe, Kirk Bryan, and Michael J. Spelman, "A Global Ocean-Atmosphere Climate Model: Part I. The Atmospheric Circulation," *Journal of Physical Oceanography* 5, no. 1 (1975): 3–29. See also Edwards, *Vast Machine*, 155.

37. S. I. Rasool and S. H. Schneider, "Atmospheric Carbon Dioxide and Aerosols: Effects of Large Increases on Global Climate," *Science* 173, no. 3992 (1971): 138–41; Syukuro Manabe and Richard T. Wetherald, "The Effects of Doubling the Carbon Dioxide Concentration on the Climate of a General Circulation Model," *Journal of Atmospheric Sciences* 32, no. 1 (1975): 3–15. See also Edwards, *Vast Machine*, 181.

38. Earth System Sciences Committee, *Earth System Science Overview: A Program for Global Change* (Washington, DC: National Aeronautics and Space Administration, 1986). See also Erik M. Conway, *Atmospheric Science at NASA: A History* (Baltimore: Johns Hopkins University Press, 2008), 225–32.

39. Thomas Rosswall, "Introduction," *Global Change Newsletter* 1 (1989): 1–2, quote on 2. See also Chunglin Kwa, "Local Ecologies and Global Science: Discourses and Strategies of the International Geosphere-Biosphere Programme," *Social Studies of Science* 35, no. 6 (2005): 923–50.

40. See J. Leggett, W. J. Pepper, and R. J. Swart, "Emissions Scenarios for IPCC: An Update," in *Climate Change 1992: The Supplementary Report to the IPCC Scientific Assessment*, ed. J. T. Houghton, B. A. Callander, and S. K. Varney (Cambridge: Cambridge University Press, 1992), 69–96. See also Bert Bolin, *A History of the Science and Politics of Climate Change: The Role of the Intergovernmental Panel on Climate Change* (Cambridge: Cambridge University Press, 2007), 91.

41. IPCC Working Group I, "The 1992 IPCC Supplement: Scientific Assessment," in Houghton, Callander, and Varney, *Climate Change 1992*, 1–22, quote on 9.

42. William D. Nordhaus, *Warming the World: Economic Models of Global Warming* (Cambridge, MA: MIT Press, 2000).

43. Study of Critical Environmental Problems, *Man's Impact on the Global Environment: Assessment and Recommendations for Action* (Cambridge, MA: MIT Press, 1970); William H. Matthews, William W. Kellogg, and G. D. Robinson, eds., *Man's Impact on the Climate* (Cam-

bridge, MA: MIT Press, 1971); Commission on Monitoring of the Scientific Committee on Problems of the Environment (SCOPE) of the International Council of Scientific Unions (ICSU), *Global Environmental Monitoring* (Stockholm: International Council of Scientific Unions, 1971). See also Howe, *Behind the Curve*, 69–76.

44. See Barbara Ward and René Dubos, *Only One Earth: The Care and Maintenance of a Small Planet* (New York: Norton, 1972), 191–220. See also Selcer, *Postwar Origins of the Global Environment*, 201–5.

45. Ward and Dubos, *Only One Earth*, 195, 220.

46. Selcer, *Postwar Origins of the Global Environment*, 206–44.

47. Howe, *Behind the Curve*, 94–95; Selcer, *Postwar Origins of the Global Environment*, 231–34.

48. Howe, *Behind the Curve*, 7–9.

49. Howe, 61–64.

50. Bolin, *History of Science and Politics of Climate Change*, 46. See also Howe, *Behind the Curve*, 147–69.

51. Kyoto Protocol to the United Nations Framework Convention on Climate Change, December 10, 1997, U.N. Doc FCCC/CP/1997/7/Add. 1, 37 I.L.M. 22 (1998).

52. Howe, *Behind the Curve*, 170–96; Bolin, *History of Science and Politics of Climate Change*, 147–62. See also J. Timmons Roberts and Bradley C. Parks, *A Climate of Injustice: Global Inequality, North–South Politics, and Climate Policy* (Cambridge, MA: MIT Press, 2006).

53. *Kyoto Protocol*, Article 12. See also Ariel Dinar, Donald F. Larson, and Shaikh M. Rahman, *The Clean Development Mechanism (CDM): An Early History of Unanticipated Outcomes* (Hackensack, NJ: World Scientific, 2013).

54. Igor Shishlov, Romain Morel, and Valentin Bellassen, "Compliance of the Parties to the Kyoto Protocol in the First Commitment Period," *Climate Policy* 16, no. 6 (2016): 768–82; Michael Le Page, "Was Kyoto Climate Deal a Success? Figures Reveal Mixed Results," *New Scientist*, June 14, 2016, https://www.newscientist.com/article/2093579-was-kyoto-climate-deal-a-success-figures-reveal-mixed-results/.

55. Jay S. Gregg, Robert J. Andres, and Gregg Marland, "China: Emissions Pattern of the World Leader in CO_2 Emissions from Fossil Fuel Consumption and Cement Production," *Geophysical Research Letters* 35, no. 8 (April 1, 2008), https://doi.org/10.1029/2007GL032887.

56. Bill McKibben, *Oil and Honey: The Education of an Unlikely Activist* (New York: Times Books / Henry Holt, 2013), 11–15.

57. James Hansen, Makiko Sato, Pushker Kharecha, David Beerling, Robert Berner, Valerie Masson-Delmotte, Mark Pagani, Maureen Raymo, Dana L. Royer, and James C. Zachos, "Target Atmospheric Carbon Dioxide: Where Should Humanity Aim?," *Open Atmospheric Science Journal* 2 (2008): 217–31, quote on 217 (in abstract).

58. Bill McKibben, "Defining Moment for Climate Change," *Amass* 13, no. 2 (2008): 12–14; Bill McKibben, "Too Hot to Handle," *Mother Jones*, November/December 2009, 32–35; Bill McKibben, "What It Will Take to Return the Globe to 350," *Tikkun*, September/October 2010, 45–49; Andrew C. Revkin, "Campaign to Reduce Carbon Dioxide Levels Picks a Number to Make a Point," *New York Times*, October 25, 2009.

59. Paul J. Crutzen and Eugene F. Stoermer, "The 'Anthropocene,'" *Global Change Newsletter* 41 (2000): 17–18. See also Will Steffen, "Commentary on Paul J. Crutzen and Eugene F. Stoermer, 'The Anthropocene' (2000)," in *The Future of Nature: Documents of Global Change*, ed. Libby Robin, Sverker Sörlin, and Paul Warde (New Haven, CT: Yale University Press, 2013), 486–90.

60. Crutzen and Stoermer, "The 'Anthropocene,'" 17.

61. Crutzen and Stoermer, 17–18.

62. Crutzen and Stoermer, 18.

63. Crutzen and Stoermer, 17. See also Will Steffen, Jacques Grinevald, Paul Crutzen, and John McNeill, "The Anthropocene: Conceptual and Historical Perspectives," *Philosophical Transactions: Mathematical, Physical and Engineering Sciences* 369, no. 1938 (2011): 842–67.

64. Maurice F. Strong, "The Stockholm Conference: Where Science and Politics Meet," *Ambio* 1, no. 3 (1972): 73–78, quote on 78.

65. Ward and Dubos, *Only One Earth*, xiii; World Commission on Environment and Development, *Our Common Future*, 26.

66. See Bill McKibben, *The End of Nature* (New York: Random House, 1989).

67. Berrien Moore III, "Sustaining Earth's Life Support Systems—the Challenge for the Next Decade and Beyond," *Global Change Newsletter* 41 (2000): 1–2.

68. Paul J. Crutzen, "My Life with O_3, NO_x, and Other YZO$_x$ Compounds (Nobel Lecture)," *Angewandte Chemie: International Edition* 35, no. 16 (1996): 1758–77.

69. Paul J. Crutzen, "Geology of Mankind," *Nature* 415, no. 6867 (2002): 23; Jan Zalasiewicz et al., "Are We Now Living in the Anthropocene?," *GSA Today* 18, no. 2 (2008): 4–8.

70. Will Steffen et al., *Global Change and the Earth System: A Planet under Pressure* (New York: Springer, 2004).

71. Steffen et al., *Global Change and the Earth System*, v.

72. W. S. Broecker, "Cooling the Tropics," *Nature* 376 (July 20, 1995): 212–13, quote on 213. See also Erik M. Conway, "Planetary Science and the 'Discovery' of Global Warming," in *Exploring the Solar System: The History and Science of Planetary Exploration*, ed. Roger D. Launius (New York: Palgrave Macmillan, 2013), 183–202, esp. 197.

73. Clive Hamilton and Jacques Grinevald, "Was the Anthropocene Anticipated?," *Anthropocene Review* 2, no. 1 (2015): 59–72.

74. O. R. Young and Will Steffen, "The Earth System: Sustaining Planetary Life-Support Systems," in *Principles of Ecosystem Stewardship: Resilience-Based Natural Resource Management in a Changing World*, ed. F. S. Chapin III, G. P. Kofinas, and C. Folke (New York: Springer, 2009), 295–315. See also Will Steffen, Paul J. Crutzen, and J. R. McNeill, "The Anthropocene: Are Humans Overwhelming the Great Forces of Nature?," *Ambio* 36, no. 8 (2007): 614–21.

75. Young and Steffen, "The Earth System," 295.

76. See Johan Rockström, Will Steffen, Kevin Noone, Åsa Persson, F. Stuart Chapin III, Eric Lambin, Timothy M. Lenton, et al., "Planetary Boundaries: Exploring the Safe Operating Space for Humanity," *Ecology and Society* 14, no. 2 (2009), article 32, http://www.ecologyandsociety.org/vol14/iss2/art32/.

77. Rockström et al., "Planetary Boundaries," quote in abstract [no page numbers].

78. Sverker Sörlin, "Reconfiguring Environmental Expertise," *Environmental Science & Policy* 28 (2013): 14–24.

79. Dipesh Chakrabarty, "The Climate of History: Four Theses," *Critical Inquiry* 35, no. 2 (2009): 197–222, quote on 201.

80. United Nations Framework Convention on Climate Change, *Report of the Conference of the Parties on Its Fifteenth Session, Held in Copenhagen from 7 to 19 December 2009*, FCCC/CP/2009/11 (March 30, 2010).

81. See, e.g., Andrew C. Revkin and John M. Broder, "A Grudging Accord in Climate Talks," *New York Times*, December 20, 2009.

82. See Clive Hamilton, *Requiem for a Species: Why We Resist the Truth about Climate Change* (Crows Nest, NSW: Allen & Unwin, 2010); and Dale Jamieson, *Reason in a Dark Time: Why the Struggle against Climate Change Failed—and What It Means for Our Future* (New York: Oxford University Press, 2014).

83. Bill McKibben, *Eaarth: Making a Life on a Tough New Planet* (New York: Time Books, 2010), xv. Some of these ideas can be found in McKibben's pre-Copenhagen work as well; see, for instance, Bill McKibben, *Deep Economy: The Wealth of Communities and the Durable Future* (New York: Times Books, 2007).

84. McKibben, *Eaarth*, 183.

85. See, e.g., David Keith, *A Case for Climate Engineering* (Cambridge, MA: MIT Press, 2013); Clive Hamilton, *Earthmasters: The Dawn of the Age of Climate Engineering* (New Haven, CT: Yale University Press, 2013). On the prehistory of geoengineering, see James Rodger Fleming, *Fixing the Sky: The Checkered History of Weather and Climate Control* (New York: Columbia University Press, 2010); Jacob Darwin Hamblin, *Arming Mother Nature: The Birth of Catastrophic Environmentalism* (New York: Oxford University Press, 2013); and Kristine C. Harper, *Make It Rain: State Control of the Atmosphere in Twentieth-Century America* (Chicago: University of Chicago Press, 2017).

86. Paul J. Crutzen, "Albedo Enhancement by Stratospheric Sulfur Injections: A Contribution to Resolve a Policy Dilemma?," *Climatic Change* 77 (2006): 211–19, quote on 217. This was not the first time Crutzen had proposed exploring the feasibility of geoengineering; see, e.g., Crutzen, "Geology of Mankind," 23.

87. Royal Society, *Geoengineering the Climate: Science, Governance and Uncertainty* (London: Royal Society, 2009), quote on ix. See also Jack Stilgoe, *Experiment Earth: Responsible Innovation in Geoengineering* (London: Routledge, Taylor & Francis Group, 2015), 103–28.

88. Stilgoe, *Experiment Earth*, 15–17. See also Brian Launder and J. Michael T. Thompson, eds., *Geo-Engineering Climate Change: Environmental Necessity or Pandora's Box?* (New York: Cambridge University Press, 2010).

89. Amy Dahan and Stefan Aykut, "After Copenhagen, Revisiting Both the Scientific and Political Framings of the Climate Change Regime," in *Global Change, Energy Issues and Regulation Policies*, ed. Jean Bernard Saulnier and Marcelo D. Varella (Dordrecht: Springer, 2013), 221–38.

90. See, e.g., Jason W. Moore, ed., *Anthropocene or Capitalocene? Nature, History, and the Crisis of Capitalism* (Oakland, CA: PM Press, 2016); Gregg Mitman, Marco Armiero, and Robert S. Emmett, eds., *Future Remains: A Cabinet of Curiosities for the Anthropocene* (Chicago: University of Chicago Press, 2018).

91. Jaskiran Dhillon and Nick Estes, eds., "Standing Rock, #NoDAPL, and Mni Wiconi," *Fieldsights*, Society for Cultural Anthropology, December 22, 2016, https://culanth.org/fieldsights/series/standing-rock-nodapl-and-mni-wiconi.

92. Sheila Watt-Cloutier, *The Right to Be Cold: One Woman's Fight to Protect the Arctic and Save the Planet from Climate Change* (Minneapolis: University of Minnesota Press, 2018). See also Candis Callison, *How Climate Change Comes to Matter: The Communal Life of Facts* (Durham, NC: Duke University Press, 2014), 49–55.

93. Urban Sustainability Directors Network, "Leading Mayors Launch International Carbon Neutral Cities Alliance," press release, March 27, 2015, https://www.usdn.org/public/page/15/Carbon-Neutral-Cities-Alliance-Press-Release-March-2015.

94. Coral Davenport, "Nations Approve Landmark Climate Accord in Paris," *New York Times*, December 12, 2015, https://www.nytimes.com/2015/12/13/world/europe/climate-change-accord-paris.html.

95. United Nations Framework Convention on Climate Change, Adoption of the Paris Agreement, FCCC/CP/2015/L.9/Rev.1, 2015, Annex (Paris Agreement) (December 12, 2015). See also Daniel Klein, María Pía Carazo, Meinhard Doelle, Jane Bulmer, and Andrew Higham, eds., *The Paris Agreement on Climate Change: Analysis and Commentary* (Oxford: Oxford University Press, 2017).

96. Bolin, *History of the Science and Politics of Climate Change*, 40.

97. *Future Earth Initial Design: Report of the Transition Team* (Paris: International Council for Science, 2013); Ninad Bondre, with contributions from Sybil Seitzinger and Wendy Broadgate, "Towards Future Earth: Evolution or Revolution?," *IGBP Global Change Magazine*, November 2015, 32–35. See also Sandra van der Hel, "New Science for Global Sustainability? The Institutionalisation of Knowledge Co-production in Future Earth," *Environmental Science & Policy* 61 (July 2016): 165–75.

98. See V. Masson-Delmotte, P. Zhai, H. O. Pörtner, D. Roberts, J. Skea, P. R. Shukla, A. Pirani, et al., eds., *Global Warming of 1.5°C: An IPCC Special Report on the Impacts of Global Warming of 1.5°C above Pre-industrial Levels and Related Global Greenhouse Gas Emission Pathways, in the Context of Strengthening the Global Response to the Threat of Climate Change, Sustainable Development, and Efforts to Eradicate Poverty* (Geneva: World Meteorological Organization, 2018).

99. Mike Hulme, "Cosmopolitan Climates: Hybridity, Foresight and Meaning," *Theory, Culture & Society* 27 (2010): 267–76; Kathryn Yusoff and Jennifer Gabrys, "Climate Change and the Imagination," *Wiley Interdisciplinary Reviews: Climate Change* 2, no. 4 (2011): 516–34; Samuel Randalls, "Climatic Globalities: Assembling the Problems of Global Climate Change," in *The Politics of Globality Since 1945: Assembling the Planet*, ed. R. van Munster and C. Sylvest (London: Routledge, 2016), 145–63.

100. John Houghton, foreword to Houghton, Jenkins, and Ephraums, *Climate Change*, v–vi, quote on vi.

CONCLUSION

1. See Bruno Latour, *Science in Action: How to Follow Scientists and Engineers through Society* (Cambridge, MA: Harvard University Press, 1987); and Bruno Latour, *Reassembling the Social: An Introduction to Actor-Network-Theory* (New York: Oxford University Press, 2005).

2. Bruno Latour, *Down to Earth: Politics in the New Climatic Regime* (Medford, MA: Polity

Press, 2018), 40–41. For a similar argument, see Jane Bennett, *Vibrant Matter: A Political Ecology of Things* (Durham, NC: Duke University Press, 2010), 102.

3. See, e.g., Donna J. Haraway, *Primate Visions: Gender, Race, and Nature in the World of Modern Science* (New York: Routledge, 1989).

4. Donna J. Haraway, "Sympoiesis: Symbiogenesis and the Lively Arts of Staying with the Trouble," in *Staying with the Trouble: Making Kin in the Chthulucene* (Durham, NC: Duke University Press, 2016), 58–98. See also Donna J. Haraway, "Tentacular Thinking," in *Staying with the Trouble*, 30–57.

5. See Tim Ingold, "Point, Line, Counterpoint: From Environment to Fluid Space," in *Being Alive: Essays on Movement, Knowledge and Description* (New York: Routledge, 2011), 76–88; Stacy Alaimo, *Bodily Natures: Science, Environment, and the Material Self* (Bloomington: Indiana University Press, 2010); Timothy Morton, *The Ecological Thought* (Cambridge, MA: Harvard University Press, 2010); and Bennett, *Vibrant Matter*.

6. Christopher Duffy et al., "CZO: Susquehanna/Shale Hills Critical Zone Observatory," US National Science Foundation, Award Abstract #0725019, https://www.nsf.gov/awardsearch/showAward?AWD_ID=0725019.

7. Charles Fergus, "The Critical Zone," *Penn State News*, September 28, 2010, https://news.psu.edu/story/141371/2010/09/28/research/critical-zone.

8. Yves Goddéris and Susan L. Brantley, "Earthcasting the Future Critical Zone," *Elementa: The Science of the Anthropocene* 1 (2013), https://doi.org/10.12952/journal.elementa.000019; S. Brantley, T. White, N. West, J. Z. Williams, B. Forsythe, D. Shapich, J. Kaye, et al., "Susquehanna Shale Hills Critical Zone Observatory: Shale Hills in the Context of Shaver's Creek Watershed," *Vadose Zone Journal* 17, no. 1 (2018), 19 pp., https://doi.org/10.2136/vzj2018.04.0092.

9. National Research Council, *Basic Research Opportunities in Earth Science* (Washington, DC: National Academies Press, 2001), 35–38.

10. NRC, *Basic Research Opportunities in the Earth Sciences*, 2. The term *critical zone* had been introduced informally several years earlier by one of the authors of the report, Gail Ashley.

11. S. L. Brantley, T. S. White, A. F. White, D. Sparks, D. Richter, K. Pregitzer, L. Derry, et al., *Frontiers in Exploration of the Critical Zone: Report of a Workshop Sponsored by the National Science Foundation*, October 24–26, 2005, Newark, Delaware (2006), 30 pp., https://criticalzone.org/images/national/associated-files/1National/frontiers_in_exploration_of_the_Critical_Zone.pdf.

12. Daniel deB. Richter Jr. and Megan L. Mobley, "Monitoring Earth's Critical Zone," *Science* 326, no. 5956 (2009): 1067–68.

13. As of early 2019, more than 230 CZOs and similar research sites had been identified by the Critical Zone Exploration Network's "CZEN Site Seeker," https://www.czen.org/site_seeker (accessed February 12, 2019). See also Li Guo and Henry Li, "Critical Zone Research and Observatories: Current Status and Future Perspectives," *Vadose Zone Journal* 15, no. 9 (2016), 14 pp., https://doi.org/10.2136/vzj2016.06.0050.

14. J. Gaillardet, I. Braud, F. Hankard, S. Anquetin, O. Bour, N. Dorfliger, J. R. de Dreuzy, et al., "OZCAR: The French Network of Critical Zone Observatories," *Vadose Zone Journal* 17, no. 1 (2018), 24 pp., https://doi.org/10.2136/vzj2018.04.0067.

15. Bruno Latour, "Some Advantages of the Notion of 'Critical Zone' for Geopolitics," *Pro-

cedia Earth and Planetary Science 10 (2014): 3–6; Alexandra Arènes, Bruno Latour, and Jérôme Gaillardet, "Giving Depth to the Surface: An Exercise in the Gaia-graphy of Critical Zones," *Anthropocene Review* 5, no. 2 (2018): 120–35.

16. James T. Callahan, "Long-Term Ecological Research," *BioScience* 34, no. 6 (1984), 363–67. See also Elena Aronova, Karen S. Baker, and Naomi Oreskes, "Big Science and Big Data in Biology: From the International Geophysical Year through the International Biological Program to the Long Term Ecological Research (LTER) Network, 1957–Present," *Historical Studies in the Natural Sciences* 40, no. 2 (2010): 183–224.

17. US National Science Foundation, *CZO: Critical Zone Observatories: US NSF National Program*, 2014, http://criticalzone.org/images/national/associated-files/1National/CZO _Brochure2_2014v6-copy.pdf.

18. See, e.g., Daniel D. Richter and Sharon A. Billings, "'One Physical System': Tansley's Ecosystem as Earth's Critical Zone," *New Phytologist* 206, no. 3 (2015): 900–912.

19. Energy Transfer Partners, "Energy Transfer Announces Crude Oil Pipeline Project Connecting Bakken Supplies to Patoka, Illinois and to Gulf Coast Markets," press release, June 25, 2014, http://ir.energytransfer.com/phoenix.zhtml?c=106094&p=irol-newsArticle&ID =1942689.

20. Nick Estes, *Our History Is the Future: Standing Rock versus the Dakota Pipeline, and the Long Tradition of Indigenous Resistance* (Brooklyn, NY: Verso, 2019).

21. Robinson Meyer, "Oil Is Flowing through the Dakota Access Pipeline," *Atlantic*, June 9, 2017, https://www.theatlantic.com/science/archive/2017/06/oil-is-flowing-through-the-dakota -access-pipeline/529707/.

22. Jaskiran Dhillon and Nick Estes, "Introduction: Standing Rock, #NoDAPL, and Mni Wiconi," *Fieldsites*, Society for Cultural Anthropology, December 22, 2016, https://culanth.org /fieldsights/1007-introduction-standing-rock-nodapl-and-mni-wiconi.

23. Edward Valandra, "We Are Blood Relatives: No to the DAPL," *Fieldsites*, Society for Cultural Anthropology, December 22, 2016, https://culanth.org/fieldsights/we-are-blood -relatives-no-to-the-dapl.

24. Kim TallBear, "Badass (Indigenous) Women Caretake Relations: #NoDAPL, #IdleNo More, #BlackLivesMatter," *Fieldsites*, Society for Cultural Anthropology, December 22, 2016, https://culanth.org/fieldsights/badass-indigenous-women-caretake-relations-no-dapl-idle-no -more-black-lives-matter; Kyle T. Mays, "From Flint to Standing Rock: The Aligned Struggles of Black and Indigenous People," *Fieldsites*, Society for Cultural Anthropology, December 22, 2016, https://culanth.org/fieldsights/from-flint-to-standing-rock-the-aligned-struggles-of -black-and-indigenous-people.

25. Anne Spice, "Interrupting Industrial and Academic Extraction on Native Land," *Fieldsites*, Society for Cultural Anthropology, December 22, 2016, https://culanth.org /fieldsights/interrupting-industrial-and-academic-extraction-on-native-land. On the image of the "ecological Indian," see J. Donald Hughes, *American Indian Ecology* (El Paso: Texas Western Press, 1983); Shepard Krech III, *The Ecological Indian: Myth and History* (New York: W. W. Norton, 1999); Michael E. Harkin and David Rich Lewis, eds., *Native Americans and the Environment: Perspectives on the Ecological Indian* (Lincoln: University of Nebraska Press, 2007); and Gregory D. Smithers, "Beyond the 'Ecological Indian': Environmental Politics and

Traditional Ecological Knowledge in Modern North America," *Environmental History* 20, no. 1 (2015): 83–111.

26. Saul Elbein, "The Youth Group That Launched a Movement at Standing Rock," *New York Times Magazine*, January 31, 2017, https://www.nytimes.com/2017/01/31/the-youth-group -that-launched-a-movement-at-standing-rock.html.

27. "A Violation of Tribal & Human Rights: Standing Rock Chair Slams Approval of Dakota Access Pipeline," *Democracy Now!*, February 8, 2017, https://www.democracynow.org/2017 /2/8/a_violation_of_tribal_human_rights.

28. Kyle Powys Whyte, "Settler Colonialism, Ecology and Environmental Injustice," *Environment & Society: Advances in Research* 9 (2018): 129–48.

29. Cassie Martin, "At COP21, Finding Hope for Climate in the 'Aerocene,'" MIT News, December 21, 2015, http://news.mit.edu/2015/cop21-finding-hope-climate-aerocene-1221.

30. Tomás Saraceno, Sasha Engelmann, and Bronislaw Szerszynski, "Becoming Aerosolar: From Solar Sculptures to Cloud Cities," in *Art in the Anthropocene: Encounters Among Aesthetics, Politics, Environments and Epistemologies*, ed. Heather Davis and Etienne Turpin (London: Open Humanities Press, 2015), 57–62; *Tomás Saraceno: Becoming Aerosolar* (Vienna: Österreichische Galerie Belvedere, 2015); Nicholas Shapiro, "Alter-Engineered Worlds," 2015, https://aerocene.org/newspaper/shapiro; Nicholas Shapiro, Jody Roberts, and Nasser Zakariya, "A Wary Alliance: From Enumerating the Environment to Inviting Apprehension," *Engaging Science, Technology, and Society* 3 (2017): 575–602, esp. 593–95.

31. Aerocene Foundation, "The Aerocene Backpack," n.d., accessed April 20, 2019, http:// aerocene.org/#explorer-beta-section.

32. Aerocene Foundation, "New Record for Fully Solar Powered Human Flight," n.d., accessed April 20, 2019, http://aerocene.org/new-record-france-human-flight.

33. Aerocene Foundation, *Manifeste Aerocene / Aerocene Manifesto*, accessed August 7, 2019, https://aerocene.org/wp-content/uploads/2019/04/Aerocene_Manifesto.pdf.

34. Some earlier forms of environmental art, such as the work of Helen and Newton Harrison beginning in the 1970s, did attempt to contribute practically to the solution of environmental problems. See James Nisbet, "Contemporary Environmental Art," in *The Routledge Companion to the Environmental Humanities*, ed. Ursula K. Heise, Jon Christensen, and Michelle Niemann (New York: Routledge, 2017), 301–12.

35. Clare McAndrew, *The Art Market 2019* (Basel: Art Basel; Zurich: UBS, 2019), 104–13.

36. Jennifer Gabrys and Kathryn Yusoff, "Arts, Sciences and Climate Change: Practices and Politics at the Threshold," *Science as Culture* 21, no. 1 (2011): 1–24; Allison Carruth, "Urban Ecologies and Social Practice Art," *Resilience: A Journal of the Environmental Humanities* 1, no. 1 (2014), https://muse.jhu.edu/article/565577; Eben Kirksey, ed., *The Multispecies Salon* (Durham, NC: Duke University Press, 2014); Davis and Turpin, eds., *Art in the Anthropocene*.

37. Sasha Hildegard Engelmann, "The Cosmological Aesthetics of Tomás Saraceno's Atmospheric Experiments" (PhD diss., University of Oxford, 2017), 254.

38. See Tim Ingold, "On Flight," in *Aerocene*, ed. Tomás Saraceno (Milan: Skira, 2017), 132–39; and Timothy Morton, "Floating as Ecological Action," in Saraceno, *Aerocene*, 140–46.

39. Aerocene, accessed April 20, 2019, http://aerocene.org/.

40. Aerocene Foundation, *Manifeste Aerocene / Aerocene Manifesto.*

41. Morton, "Floating as Ecological Action," 141.

42. Deborah Bird Rose and Libby Robin, "The Ecological Humanities in Action: An Invitation," *Australian Humanities Review* 31–32 (2004), http://australianhumanitiesreview.org/2004/04/01/the-ecological-humanities-in-action-an-invitation.

43. Deborah Bird Rose, Thom van Dooren, Matthew Chrulew, Stuart Cooke, Matthew Kearnes, and Emily O'Gorman, "Thinking Through the Environment, Unsettling the Humanities," *Environmental Humanities* 1 (2012): 1–5, quote on 1, https://environmentalhumanities.org/arch/vol1/EH1.1.pdf; "About the EHL," KTH Royal Institute of Technology, accessed April 20, 2019, https://www.kth.se/en/abe/inst/philhist/historia/ehl/about-the-ehl.

44. Illinois Program for Research in the Humanities, "Illinois Program for Research in the Humanities Awarded $2,050,000 by Andrew W. Mellon Foundation to Support New Humanities Initiatives," press release, January 6, 2015, https://blogs.illinois.edu/view/8303/759148; Penn Arts & Sciences, "Penn Program in Environmental Humanities Receives Mellon Support," *Omnia*, September 26, 2017, https://omnia.sas.upenn.edu/story/penn-program-environmental-humanities-receives-mellon-support; Emily Halnan, "New Grant Bolsters UO Center for Environmental Humanities," University of Oregon, *Around the O*, April 4, 2018, https://around.uoregon.edu/content/new-grant-bolsters-uo-center-environmental-humanities.

45. Rose et al., "Thinking Through the Environment"; Ursula K. Heise and Allison Carruth, "Introduction to Focus: Environmental Humanities," *American Book Review* 32, no. 1 (2011): 3; Sverker Sörlin, "Environmental Humanities: Why Should Biologists Interested in the Environment Take the Humanities Seriously?," *BioScience* 62, no. 9 (2012): 788–89; Hannes Bergthaller, Rob Emmett, Adeline Johns-Putra, Agnes Kneitz, Susanna Lidström, Shane McCorristine, Isabel Pérez Ramos, Dana Phillips, Kate Rigby, and Libby Robin, "Mapping Common Ground: Ecocriticism, Environmental History, and the Environmental Humanities," *Environmental Humanities* 5 (2014): 261–76; Ursula K. Heise, "Introduction: Planet, Species, Justice—and the Stories We Tell about Them," in *Routledge Companion to the Environmental Humanities*, ed. Heise, Christensen, and Niemann, 1–10; Robert S. Emmett and David E. Nye, *The Environmental Humanities: A Critical Introduction* (Cambridge, MA: MIT Press, 2017), 1–22.

46. Sörlin, "Environmental Humanities," 788. See also Gísli Pálsson, Bronislaw Szerszynski, Sverker Sörlin, John Marks, Bernard Avril, Carole Crumley, Heide Hackmann, et al., "Reconceptualizing the 'Anthropos' in the Anthropocene: Integrating the Social Sciences and Humanities in Global Environmental Change Research," *Environmental Science & Policy* 28 (2013): 3–13.

47. Noel Castree, William M. Adams, John Barry, Daniel Brockington, Bram Büscher, Esteve Corbera, David Demeritt, et al., "Changing the Intellectual Climate," *Nature Climate Change* 4 (2014): 763–68.

48. Stephanie LeMenager, "The Environmental Humanities and Public Writing: An Interview with Rob Nixon," *Resilience: A Journal of the Environmental Humanities* 1, no. 2 (2014), 12 pp., https://www.jstor.org/stable/10.5250/resilience.1.2.006.

INDEX

Page numbers in italics refer to figures.